DIGITAL OIL

INFRASTRUCTURES SERIES

Edited by Geoffrey C. Bowker and Paul N. Edwards

A list of books in the series appears at the back of the book.

DIGITAL OIL

Machineries of Knowing

ERIC MONTEIRO

The MIT Press
Cambridge, Massachusetts
London, England

The MIT Press would like to thank the anonymous peer reviewers who provided comments on drafts of this book. The generous work of academic experts is essential for establishing the authority and quality of our publications. We acknowledge with gratitude the contributions of these otherwise uncredited readers.

This book was set in Adobe Garamond and Berthold Akzidenz Grotesk by Jen Jackowitz. Printed and bound in the United States of America.

Library of Congress Cataloging-in-Publication Data

Names: Monteiro, Eric, author.
Title: Digital oil : machineries of knowing / Eric Monteiro.
Description: Cambridge, Massachusetts : The MIT Press, [2022] |
    Series: Infrastructures. | Includes bibliographical references and index.
Identifiers: LCCN 2022003259 (print) | LCCN 2022003260 (ebook) |
    ISBN 9780262544672 (paperback) | ISBN 9780262372282 (pdf) |
    ISBN 9780262372299 (epub)
Subjects: LCSH: Petroleum industry and trade—Norway. | Oil fields—Norway—Data
    processing. | Oil field equipment and supplies industry—Norway—Technological
    innovations.
Classification: LCC HD9575.N62 M66 2022 (print) | LCC HD9575.N62 (ebook) |
    DDC 338.2/72809481—dc23/eng/20220124
LC record available at https://lccn.loc.gov/2022003259
LC ebook record available at https://lccn.loc.gov/2022003260

10   9   8   7   6   5   4   3   2   1

Contents

III IMPLICATIONS

Preface

This book represents a long journey. It mirrors travels that make up much of my academic biography. Ostensibly, with a chair in information systems in a computer science department at an engineering-dominated university, I have developed perspectives in this book that engagements along this journey have shaped in important ways. I dwell on a select few to create a backdrop for what follows.

Trained as an engineer, I was drawn to logic for my graduate studies at the intersection of the humanities and informatics at the University of Oslo. Crucially, however, this interest was tied to logic as a language, not a purely technical discipline. Particularly influential were Husserl's formulations of *constructive* mathematics and logic, based on his phenomenological perspectives. At the Norwegian Computing Centre, Oslo, I found myself in the middle of critical, socially informed discourses on the conditions, manifestations, and consequences of Scandinavian-based participatory modes of technology development. Disciplinary boundaries were porous. The field of science and technology studies (STS) had a formative influence on me, first through the Centre for Technology, Innovation and Culture, Oslo, and then at the Centre for Technology and Society, Trondheim. Alongside a theoretical curiosity about STS, I developed a growing empirical interest in large-scale (*infrastructure*) technology efforts with implicated standardization as, seemingly, this went beyond the existing participatory methods for technology development. Relocating to the Norwegian University of Science and Technology, Trondheim, I became attracted to the perspectives of—and, not least, experiences with—the politics of participatory and interventionist forms of organizational change being pursued at the Department of Industrial Economics and Technology Management. Cultural perspectives on standardization, objectification, and quantification out of the Department of Anthropology were important in broadening my notion of standardization.

My academic coming of age, then, is the result of stitching together a network of colleagues and collaborators from a variety of disciplines and camps. It has been driven by the instinct to challenge my own intellectual comfort zone, wary of growing too comfortable in any one place.

In a final comment on the theme of data science and artificial intelligence (AI) emerging in this book, I must admit that I had no intention whatsoever of revisiting AI; I had my fill a couple of decades ago. But recent demand from external partners from the private and public sectors in research projects—within oil, as I report from here, but also from my research stream in health care—nudged me toward the theme of datafication and data-driven approaches. Having spent much of my professional career explaining why various technology efforts had failed, I was intrigued by how data-driven data science, for particular purposes, apparently "works" in ways beyond what we have presently accounted for theoretically. In short, my curiosity was stirred by a "works in practice, not in theory" situation.

Trondheim, Norway, Autumn 2021

Acknowledgments

I have benefited greatly from interactions with a lot of people over the long period I have pondered and explored parts of this book. Providing an exhaustive list is prohibitive.

A number of close colleagues provided invaluable comments based on snippets and arguments, small and large, throughout the process of developing this book. With or without realizing it, their conversations supplied much-needed sparring about early ideas: Petter Almklov, Michael Barrett, Bendik Bygstad, Samer Faraj, Ole Hanseth, Vidar Hepsø, Jannis Kallinikos, Neil Pollock, Knut H. Rolland, Susan Scott, Georg von Krogh, and Robin Williams.

Several scholars contributed to my research during the seminars, visits, and events I attended in locations including Ascona, Barcelona, Cambridge, Copenhagen, Edinburgh, Oslo, London, Seattle, Trondheim, Umeå, Warwick, and Zurich: Christina Alaimo, Panos Constantinides, Ola Henfridsson, Jonny Holmström, James Howison, Steve Jackson, Alexander Kempton, Davide Nicolini, David Ribes, Susan Scott, Geoff Walsham, and Youngjin Yoo. By challenging my arguments, they helped clarify otherwise muddled thinking.

Many colleagues have provided inspiration, energy, and indirect support in ways I struggle to account for adequately: Margunn Aanestad, Jørn Braa, Kristin Braa, Gunnar Ellingsen, Øystein Fossen, Morten Hatling, Roger Klev, Tord Larsen, Morten Levin, Emil Røyrvik, Jens Røyrvik, Sundeep Sahay, Knut H. Sørensen, and Arild Waaler.

I am deeply indebted to Marius Mikalsen, Elena Parmiggiani, and Thomas Østerlie. They are coauthors of the series of published papers that make up the point of departure for the chapters in part II of this book. In reworking and molding these articles to fit within the book's broader arguments, I have been in dialogue with and responded to feedback from Marius, Elena, and Thomas. Thus, it is reasonable to acknowledge their role in the respective chapters in part II as "written with" me.

I also want to thank the coauthors of other papers related to the arguments put forward here: Petter Almklov, Vidar Hepsø, Gasparas Jarulaitis, and Knut H. Rolland.

The book contains a number of figures and images, without which much would be lost. I thank these companies, agencies, institutions, colleagues, and open data repositories for granting permission to reproduce: Norsk Oljemuseum (Shadé B. Martins), Equinor, Irina Pene, Norne open data set, Norsk Olje og Gass, Norwegian Petroleum Directorate, Elena Parmiggiani, Shell, TechnipFMC, MAREANO/Institute of Marine Research, and Volve open data set.

I am grateful for help from the MIT Press, not least for the formative comments from reviewers. I also want to thank Geof Bowker and Paul Edwards for support and encouragement throughout the years. Justin Kehoe was a great help in navigating the final hurdles at the publisher.

Finally, I am thankful for the patience, space, and support granted by my family throughout this whole journey.

Part of the research in this book has been supported by Research Council of Norway grants 163365 Aksio, 213115/O70 Digital Oil, and 237898/O30 Centre for Research-Based Innovation (SFI) Sirius.

1 INTRODUCTION

Action and knowing are *situated*. Coined more than thirty years ago (Suchman 1987), the situated nature of our engagement with digital technologies has shaped many of the socially informed empirical accounts. Early and influentially, Zuboff (1988) studied the digitalization of work in industrial and office settings. She underscored the tactile and embodied competence of predigital daily work routines. For instance, in her study of an industrial pulp mill she emphasized the importance of the operators dipping their fingers into the pulp and tasting (!), smelling, and feeling the temperature and texture of the pulp in order to competently engage in the everyday running of the mill. In a similar vein, practice theory–based accounts underscore how our engagement with digital technologies is "*emergent* (arising from everyday activities and thus always 'in the making'), *embodied* (as evident in such notions as tacit knowing and experimental learning), and *embedded* (grounded in the situated socio-historic contexts of our lives and work" (Orlikowski 2006, 460; emphasis in the original). Hence, knowing is "a situated knowing constituted by a person acting in a particular setting and engaging aspects of the self, the body, and the physical and social worlds" (Orlikowski 2002, 252). Within the industrial sector empirically examined in this book, offshore oil and gas exploration and production, the typical image of an operator is a roughneck: smeared in oil and grease, hard hat on his (not "her"!) head, wrenching loose a 31.6-foot drill pipe.

A broadly practice-oriented perspective—underscoring qualitative, embodied, situated action—has been highly influential and underpins many of the critical, socially informed studies of work and technology. It resonates deeply with my own perspective. However, I do have issues with what such a perspective risks leaving out. In this book I analyze digitalization as ongoing attempts, regularly met with opposition and setbacks, to quantify the qualitative. In other words, I explore whether the above outlined practice-based perspective might have overstated the role and scope of the qualitative in present-day digitally enabled practices of knowing (see figure 1.1).

Empirically, the transformation of work practices has been in full swing for quite some time in the oil and gas industry (Autor 2015; Thune et al. 2018). Roughnecks are increasingly rare. The majority of hydrocarbons produced on the Norwegian continental shelf is by subsea production facilities residing on the bottom of the sea, untouched, as it were, by human hands and remotely operated from onshore control rooms based on real-time sensor streams measuring temperature, pressure, and volume.

The digitally enabled transformations in offshore oil and gas of work practices, roles, and organization that I analyze in this book are not the result of radical, discontinuous change. This book does not promote the imagery of a great digital divide between, on the one hand, early or predigital practices and, on the other hand, more recent forms of digitalization with their emphasis on intelligent systems (artificial intelligence, or AI), blockchain, digital platforms, social media, and the Internet of Things (IoT) as found in more managerial strands of digitalization (cf. Brynjolfsson and McAfee 2014; McAfee et al. 2012; Davenport 2014). On the contrary, this book firmly subscribes to a perspective of technological change emphasizing evolutionary, small-step, socially negotiated change; ongoing experimentation regularly meeting with setbacks; and opposition or outright failure. Still, the cumulative changes over the last couple of decades, the time frame of this book, have significantly changed work routines and roles. The relevance and significance of prominent forms of digitalization during this period, notably that of IoT and data-driven approaches, is not their novelty per se (new digital technologies come and go) but the way they allow attempts (again, subject to negation, conflict, and opposition) at quantifying other kinds of tasks imbued with qualitative and/or tactile qualities. In other words, a fascination with the ongoing attempts at pressing the limits or scope of digitalization

Figure 1.1
Work practices in oil: traditional roughneck (*top*) versus control room based (*bottom*).
Source: Reproduced by permission from Husmo Foto/Norsk Oljemuseum and Shadé B. Martins/Norsk Oljemuseum, respectively.

animates my analysis. The ambition of this book, then, is to balance a healthy skepticism of proclamations for revolutionary or radical change against an empirical, phenomenon-oriented openness to interestingly different aspects of the new in the old. To that end, a rough historic outline of digitalization within the industrial context under study is helpful.

Digitalization of the Norwegian offshore industry started modestly in the 1980s and early 1990s. The emphasis was on largely stand-alone process control systems. Picking up speed through the 1990s and 2000s, digitalization efforts shifted to enabling data communication, notably between on- and offshore operators, as well as access to sensor data (including downhole). These efforts, known alternatively as *eField*, *intelligent fields*, or *integrated operations* (Rosendahl and Hepsø 2013), challenged existing organizational routines and division of labor. During the last couple of decades, offshore personnel and tasks have been shifted onshore. Videoconferencing, email, and instant messaging are frequently used for communication and collaboration between offshore installations and the mainland. With the availability of real-time sensor data and new engineering applications for visualizing and manipulating this data, onshore engineers can also actively participate in monitoring, diagnosing, and controlling offshore processes (see figure 1.2).

What this implies is that current, everyday offshore oil and gas work is significantly and qualitatively different from the outlined practice-oriented position with its emphasis on the tactile, embodied work characteristic of practices twenty-five years ago. Not in a radical move, but evolutionarily and cumulatively over a couple of decades, the content and context of work practices have profoundly changed. An offshore oil platform off Norway with its subsequently connected subsidiary production platforms as well as unmanned subsea facilities is today a massively instrumented production facility. For instance, the Ekofisk field in the southwestern part of the North Sea has about ten thousand real-time sensor feeds from different depths down in the well and from the topside, valves, processing, and transportation equipment. Sensor-based, IoT-rendered "reality" is a sine qua non for the work practices constituting everyday offshore operations. Compounded by a sharp drop in oil prices in June 2014, automating and/or shifting more and more functions to digitally enabled, remote operations is a chronic cry.

Figure 1.2
Subsea installation for oil production.
Source: Reproduced by permission from TechnipFMC and Shell.

For instance, an actively pursued vision is that of the unmanned *subsea factory*. Until recently, oil and gas production on the Norwegian continental shelf has involved processing facilities, pumps, separators, and compressors located on floating platforms to inject water and gas into the geological reservoir. This is needed as, after an initial period of sufficient pressure in the reservoir, the production of hydrocarbons gradually releases and hence decreases the reservoir's pressure. Left unchecked, a significant portion (more than 70–80 percent) of the oil and gas reserves could not be "sucked" up to the platform. After the initial period, when the reservoir pressure will push hydrocarbons to the surface only when punctured by a drill string, the pressure must be maintained during production by injecting gas or water into the reservoir. In this manner, oil recovery may be increased to 50 percent of the oil reserve of the reservoir, from a typical global rate of 25–30 percent. The processing facilities, separators, injectors, compressors, and pumps required for this must fight for the precious little space available on a floating platform. They are energy consuming (hence polluting). The vision of the subsea factory, then, is to move away from this situation. The ongoing aim is to fully automate the process, relying on electrical rather than today's fossil power.[1] In parallel, existing manned platforms are increasingly being operated remotely, as announced by the dominant operator, Statoil (2017; now named Equinor), in 2017 and reports in the news (Johansen and Kristensen 2017; see figure 1.3 depicting a compressor component).

Accordingly, the issue in offshore oil and gas is not so much one of whether to substitute the manual, tactile, and embodied work practices of roughnecks with IoT-rendered, digitally enabled ones. The issue is one of sequence, scope, and pace. This transformation from manual to automated and/or remotely operated operations, however, is gradual and negotiated at every step along the trajectory. It is subject to contestation from unions and national safety authorities. The ongoing efforts of automation are hardly expressions of technological determinism. For instance, one union leader commenting on the reduction of offshore employees resulting from automation argues that they "request more compelling documentation for the consequences for safety," suggesting

Figure 1.3
Compressor component for subsea factory.
Source: Reproduced by permission from TechnipFMC and Shell.

that automation (with downsizing) gives rise to an "increased risk of accidents" (Fredriksen et al. 2016). Likewise, in 2018 Norway's Petroleum Safety Authority (Petroleumstilsynet [Ptil]; Ptil 2018) issued a report pointing out the potential hazards to health, safety, and the environment from ongoing and future digitalization. The empirical focus on offshore oil and gas provides an entry point to engage with and challenge ongoing conceptualizations of digitalization. This book is a sustained, conceptually oriented, empirically grounded analysis of digitalization within the industrial setting of commercial oil and gas. With hydrocarbons on the Norwegian continental shelf located kilometers below the seabed, work practices and operational decision-making rely on IoT-based data from seismic (acoustic reflections of subsurface rocks), well logging (electromagnetic and radioactive measurement of rock properties along an oil well), and real-time production data (measurements of flow volume, temperature, pressure, and chemical composition). Resulting from the cumulative evolution of changes in work and technology over the last couple of decades, this book is distinctly different from images of roughnecks smeared in oil. It is a case of the industrial IoT. It is not a portrait of a distant future, as more than half of the oil and gas produced today comes from unmanned, sensor-monitored, data-driven subsea facilities. In short, this book is about the datafication in oil; it is about *digital* oil.

The first and fundamental step in developing the theoretical position informing this book is to address the still-dominant dichotomous separation between, on the one hand, the physical and real and, on the other hand, the digital and "merely" virtual. To analyze the expanding reach and scope of digitalization, we need to recognize how and why this dichotomy is fundamentally flawed. As Boellstorff (2016) notes, a "staggeringly large number" of scholars rely on this dichotomy: "Much more than slips of the conceptual tongue, these conflations reflect deep-seated assumptions about value, legitimacy, and consequence [that] forecloses comprehensively examining world makings and social constructions of reality in a digital age" (387–388). The title of this book—*Digital Oil*—is based on systematically dismantling the dichotomy between the physical/real versus the virtual/digital. The challenge well into the twenty-first century is rather to "understand precisely how the digital can be real" (Boellstorff 2016, 388). In dismissing the dichotomy,

I lean on arguments made by other scholars and review a selection of the salient ones below.

A reasonable starting point for discourse on the physical/real versus the digital/virtual is Zuboff's (1988) pioneering work. She was among the first to analyze how digital technologies, not only technology in general, transform work. She coined the terms "automate" and "informate" to capture the distinction between technology in general and digital technology specifically. In her analysis all technologies, including digital, come with the potential to automate. What makes digital technologies distinct is their additional potential to informate. The defining quality that Zuboff built into her notion of informate was the insight that digital technologies do not consume their *input factors*—that is, data. Similarly, the output from digital technologies (data) may also be used as input for open-ended purposes without consuming the input factor. Informate thus ties the specific capacity of digital technologies to their ability to re-present data indefinitely and, as it were, at no cost. Thus, Zuboff made the nature and challenge of digital re-presentation of the physical world a fundamental theme.

However, she warned about the dangers of a digitally rendered reality. Like the empirical context of this book, Zuboff studied the safety-critical operation of running a large process plant, a pulp mill. In transforming from experience-based, embodied, tactile handcraft—smelling, tasting, and feeling the temperature of the pulp—to a remotely operated, digitally enabled control room, she noted the unease stemming from "digital [representations] replacing a concrete reality" (63) and how digital representations "replace the sense of hands-on" (65) and seek to "invent ways to conquer the felt distance of the referential function."

Zuboff's analysis is of equal, if not greater, relevance today as "it may have an even stronger story to tell now than it did when first published" (Burton-Jones 2014, 72) as a result of the expanding scope and reach of digitalization. However, her argument that a lack of sensory feedback undermines knowledge-based work reinforces rather than dismantles the physical/real versus digital/virtual dichotomy. This leads us to scholars who, drawing on Zuboff (1988), go on to challenge that dichotomy.

One helpful attempt is Knorr Cetina's (2009) work—specifically, her notion of a synthetic situation. The fundamental, compelling insight of situated action is that action is not determined by design or constraints. Action is contingent on local circumstances and resources. Suchman (2006; emphasis added) stipulates that by "situated actions I mean simply actions taken in the context of the *particular, concrete circumstances*" (25–26). Where unclarity, disagreement, and debate start is when it comes to detailing the exact meaning of "particular, concrete circumstances." The notion of a synthetic situation grants the same status to the physical/real and the virtual/digital with regard to their role in practices of knowing. In her case, Knorr Cetina (2009) portrays the transformation of financial trading from physically colocated settings to distributed and electronically mediated settings. As an empirical phenomenon, "a 'situation' invariably includes, and may in fact be entirely constituted by, on-screen projections" (65). The synthetic situation "stitches together an analytically constituted world made up of 'everything' potentially relevant to the interaction" (66). Empirically including digital representations is fundamentally different from the symbolic interactionalist's definition of a situation, Knorr Cetina argues, which, "despite nods . . . was, at its core . . . a physical setting or place" (63).

What synthetic situations teach us, then, is that digital representations may be as real as the physical in material knowing. Digital representations, so much more than mere representations, may under enabling social, political, and technological conditions have organizational and societal consequences and significance (Burton-Jones 2014; Kallinikos 2007; Borgmann 1999; Lusch and Nambisan 2015; Boellstorff 2016).

Sidestepping the increasingly esoteric debate on the ontological status of the digital within the debates on sociomateriality (Barad 2003; Cecez-Kecmanovic et al. 2014; Orlikowski and Scott 2008), my interest lies firmly in the epistemic implications of digitalization—that is, how digital representations are implicated in knowledge-based work practices. I am, much like Latour's (1999) instincts, interested in how knowing is *done* when, for all practical purposes, work practices increasingly require digital representations to "stand in for reality" (Bailey et al. 2012) rather than concern for what they really are (Baudrillard 1994; Ihde 1995; Baskerville et al. 2020). When Latour (1992), in one of his many pedagogic examples, likens a speed bump to a police officer, the interesting question is not whether they are the same (they obviously are not) but rather if—for the purpose of understanding the behavior of car drivers—they fill the same role in influencing car drivers' behavior regarding speeding.

The empirical focus in this book is, accordingly, a sampled set of knowledge-based work routines, with a spotlight on what the practitioners do, why they do it, what strategies they employ to accumulate evidence and credibility for what they know, and how the digitally rendered reality is interrogated.

A defining quality of digitalization—indeed, the source of its potential for radical recombination and disruption (Henfridsson et al. 2018)—is its capacity for disembedding digital representations from their originating physical objects, qualities, or processes. This capacity is what underpins the dismantling of the physical/real versus digital/virtual dichotomy as it opens up for digitally capturing what was previously the exclusive realm of the physical/real. This disembedding, or *liquefaction*, captures through digital representation what originated "from its related physical form or device" (Lusch and Nambisan 2015, 160). Liberated from the physical referent (i.e., physical object, quality, or process), digital representations may be aggregated, algorithmically manipulated, and visualized to open-endedly support the action and decision-making underpinning work practices.[2] This entails that the objects of knowing are increasingly *self-referential* digital representations rather than physical objects or processes (Kallinikos 2007). The object of knowing, increasingly, is an "algorithmic phenomenon" (Orlikowski and Scott 2016).

Another way of formulating this is by challenging our notion of what "data" are. As anything but naively representing a given reality (Jones 2019), data are highly constructed, evolving achievements (Kallinikos et al. 2013; Alaimo and Kallinikos 2020; Baskerville et al. 2020). Not necessarily a faithful representation of a given phenomenon (Burton-Jones and Grange 2013), data may be sliced, aggregated, and algorithmically manipulated because data are "'footprints' of [physical] events, rather than . . . the events themselves"

(Knorr Cetina 1999, 41). The tendency I pursue in this book, then, is that data, increasingly, *are* the phenomenon.

In this context the IoT takes on a particularly pregnant meaning. Sensors are vehicles for liquefaction: they generate digital representation inferred from physical objects, properties, and processes (Oreskes et al. 1994). Sensors, quintessential generators of liquefaction, expand in reach and scope the type of phenomenon made synthetically knowable. With this expansion in scope and reach, sensors increasingly mimic embodied perception (seeing, hearing, tactile sensation, smelling, and movement/balance; see Singh et al. 2014). Understood phenomenologically, sensors allow for sensing the life-world in already interesting but rapidly expanding richness. This gives an ironic twist to Zuboff's (1988) original argument. For her, digital technologies' lack of ability to grasp tactile qualities was the reason for the limitations in digitally enabled knowing. Humans, not digital technologies, are capable of knowing, Zuboff (1988) argues, as "I see, I touch, I smell, I hear; therefore, I know" (62). With the expanding scope of sensors, this limitation no longer holds in the same way. Clearly, all representations, not only digital ones, rely on disembedding the physical from the representation (Oreskes et al. 1994). All representations, including but certainly not limited to digital representations, re-present (Wood and Fels 1992). Digital representations are especially interesting because of the versatility and ease of their subsequent open-ended manipulation, as Zuboff's notion of informate underscores.

The implicated substitution of quality for quantity in digitalization qua liquefaction is but a potential. Technological determinism, in any shape or form, is flawed. Historical accounts teach instructive lessons about the invariable blood, sweat, and tears accompanying transformations of quality for quantity. The transition, history reminds us, is littered with setbacks and is anything but frictionless. Crosby (1997), for instance, describes how temperature, historically a qualitative phenomenon of "hot" and "uncomfortable," only slowly and gradually got replaced by quantitative measures into degrees centigrade and Fahrenheit. He also describes how quantification in one area (temperature) led to inflated expectations about what could reasonably be quantified (such as grace and virtue).

With digitalization potentially, but far from automatically, transforming work practices, the key issue in this book is how, if at all, digital representations *become organizationally real* (cf. Bailey et al. 2012; Leonardi 2012). I am concerned with the mechanisms and conditions necessary for organizational "real" digital representations to become institutionalized work practices.

Easily mistaken for a purely philosophical concern, liquefaction, or data's representational capacity—that is, data's capacity to represent a phenomenon (Zuboff 1988; Burton-Jones 2014; Kallinikos 2007; Borgmann 1999)—is increasingly recognized as being at the core of discourses over big data, data science, and data-driven machine learning (Alaimo and Kallinikos 2020; Markus 2017; Zuboff 2019). Vast, heterogeneous, and interconnected data sets—the machineries or infrastructure of knowing—are redefining the boundaries of data-driven action and decision-making. During the previous wave of AI in the 1980s–1990s, language translation was identified as the acid test for the *I* in AI, intelligence: mastering automated language translation, the argument went, was on par with full-fledged intelligence. However, Google Translate, especially after 2016 when changing to the data-driven, deep learning–based algorithm, works in ways earlier symbolic AI never did. Error prone, imperfect, and hardly producing poetry, automated language translation today works for a number of practical tasks, including reading hotel reviews before planning a summer vacation or skimming through a work-related technical report. Some scholars remain unimpressed, arguing that essentially nothing has changed: the fundamental human capacity for knowing remains beyond the scope of computerization (Dreyfus and Dreyfus 2000; Autor 2015). Others, invigorated by the potential of data science, proclaim the coming of a new paradigm in knowing in general and science in particular. Visions of data-driven knowing feed proclamations of a *fourth paradigm* of science (see Kitchin 2014). Purely inductive, data-driven predictions are to replace models and theory, creating as it were an empiricist's ideal (McAfee et al. 2012; Lazer et al. 2009; Anderson 2008; Davenport 2014).

This book pursues a middle ground between utopian exclamations of a fourth paradigm and dismissals of "business as usual" (van den Broek et al. 2021). Differing from ideologically poised positions, the mechanisms, limits, and consequences of digital transformation remain empirically open. Why then is the present version of digitalization worthy of scholarly attention? That is, how, if at all, are present manifestations of digitally mediated knowing different?

Empirically interesting, theoretically underarticulated changes have occurred in how knowledge workers in practice work with big data, interrogate their robustness, and engage in *trading zones* to provisionally reach enough of a consensus to know what to do next. In this perspective, data-driven practices of knowing are not a distant future delegated to machine learning–based predictive algorithms but are already on display. Practices of knowing in contexts saturated by vast, heterogeneous, and uncertain data—like the one I am analyzing on offshore digital oil—are already "data driven." There is, as von Krogh (2018) points out, an evolving, sufficiently different empirical, organizational phenomenon of digitalization that needs to be met with phenomenon-based theorizing.

The subtitle of this book, *Machineries of Knowing*, is an acknowledgment of addressing the transformative capacity of digitalization with an infrastructure-, not artifact-centric, lens (Borgman et al. 2013; Hanseth et al. 1996; Monteiro et al. 2013). This is the foundational theoretical premise of the knowledge infrastructure perspective that (explicit or implicit) informs the present MIT Press Infrastructures series of books. My book, with its actor-network theory affinities, is in line with such a premise, which Latour (1999) succinctly formulates as "B-52s do not fly, the U.S. Air Force flies" (182). A vivid illustration of such a perspective is provided in Ribes and Polk's (2015) account of the forty-year evolution of the knowledge infrastructure underpinning HIV/AIDS research. It illustrates the extensive network of practices (campaigns to keep subjects in the cohort motivated to participate), technologies (measurement devices, databases), and institutions.

The basic premise of an infrastructure perspective is not only confined to (knowledge) infrastructure scholars but also underscored by other critical, social theorists. This book is thus compatible with a broad church of infrastructure sensitivities and is not confined to those adhering only to knowledge infrastructures. As Morgan (2010) illustrates in the relevant context of data, be it from sensors or other sources, data "depend upon systems, conventions, authorities and all sorts of good companions to get [data] to travel well" (4).

A recurring theme in infrastructure studies is that of *inertia*—that is, how over time and as a result of entanglement in the network of practices, tools, and institutions illustrated above, infrastructures become increasingly resistant to abrupt or radical change (cf. the notion from network economics of path dependency). In Star's (1999, 381–382) early characterization of infrastructures, about half of the defining qualities tapped into notions of inertia (notably "embeddedness," "linked with conventions of practice," and "built on an installed base"). Within digitalization, the notion, notoriously ambiguous (Gillespie 2010), of a digital *platform* has recently attracted attention. Conceptualized alternatively as a (double-sided) market (Gawer 2011), a technical architecture (Tiwana 2013), an organizational form (Kornberger et al. 2017), or a business model (Parker and Alstyne 2014), the relevance of the notion of a platform in my infrastructure-based analysis of practices of knowing in digital oil is in its potential to augment the well-rehearsed arguments in infrastructure studies of inertia with interesting possibilities and mechanisms for change. There is one additional, important reason for thematizing an infrastructure perspective through the lens and vocabulary of digital platforms. Digital platforms supplement the unit of analysis of digitally enabled change from predominantly within to across organizations (see Vial 2019); that is, they promote a growing focus on sector- or industry-wide changes to complete ecosystems.[3] In addition, the emphasis on infrastructure/platforms highlights the significant shift of focus from those internal to the organization (which dominate to date) to a growing awareness of changes in the whole ecosystem of industries or domains.

With several decades of scholarly attention to computerization (Kling 1996), virtualization (Bailey et al. 2012), organizational implementation of ICT (information and communications technology; Orlikowski 2002), and social informatics (Kling 2000), to mention but a few of the labels, one might rightly wonder what, if any, is new with the proliferation of the label *digitalization*. Is digitalization but a rhetorically (and commercially) motivated relabeling of well-rehearsed arguments and insights from these preceding decades? The label of digitalization might be new, but the underlying phenomenon—the uptake of digital technologies into social practices—certainly is not. Hence, what is it and what does it entail? What, if anything, is new with digitalization?

Inflating and/or exaggerating the novelty of new concepts—underscoring discontinuities, downplaying continuities—is a well-rehearsed rhetorical strategy (Abrahamson 1991). Several vital insights and concepts have been generated from studies prior to the more recent coining of the label of digitalization. Notable ones include the situated nature of digitally enabled action (Suchman 1987); evolutionary rather than revolutionary changes (Barley 1986); side effects (complementarities) as significant as intended outcomes (Ash et al. 2004); the inclination to exaggerate short-term and underestimate long-term outcomes (Brynjulfsson and Hitt 2000); and change emerging from accumulated and concerted rather than isolated efforts (Morgan 2010).

This book promotes a particular view of digitalization. I pay due respect to the insights gained over several decades from the socially informed, critical studies of digitalization indicated above. It is firmly based in a tradition that underscores continuities over discontinuities, gradual over radical change. At the same time, I argue that three interesting aspects of digitalization are sufficiently different to warrant our scrutiny. I emphasize three characteristics of the shift in knowledge-based practices under the banner of digitalization that I believe are undertheorized, hence the focus of this book.

Objects of knowing: Data   The objects of knowing (here, geological reservoirs, oil wells, sand in produced flows of hydrocarbon, coral reefs, and fish) are increasingly digital representations. Sensors are making inroads into the qualitative in poorly understood ways and hence mimic previously perceptive or tactile qualities. However, the capacity to liquefy or disembed digital representations from their originating referent is exactly that—a capacity. The empirically grounded, theoretical concern is to characterize the circumstances underpinning digital representations actually, not merely potentially, being woven into everyday knowing practices.

Modes of knowing: Algorithmic and data driven[4]   The means and methods of knowing relevant objects are predominantly mediated digitally through a variety of tools. The quantification implicated in data-driven, machine learning–based approaches are making inroads into the heartland of the qualitative: geological interpretations, professional judgments, risk assessments, and evaluations (von Krogh 2018). Again, the focus is not so much the potential for data-driven quantification but the characterization of the circumstances underpinning "working" arrangements for particular purposes and situations.

Machineries of knowing: Infrastructure and platforms   A prominent, if not dominant, trait of previous studies of digitalization (with all its connotations and labels) is an overemphasis on the local, typically in the form of a single-site case study in one organization (Williams and Pollock 2012), and an overemphasis on singular artifacts. The research stream on knowledge or information infrastructure provides a healthy supplement by insisting on always studying local dynamics around singular artifacts in conjunction with the broader institutional, historic fabric (Borgman et al. 2013; Larkin 2013; Monteiro et al. 2013). Drawing on an infrastructure perspective, this book thus documents how digitalization processes invariably are caught up in and presuppose broader enabling circumstances. Moreover, I pursue an agenda of infrastructure studies that see digital transformation as resulting from platform-enabled ecosystems of technology, institutions, and work practices (Plantin et al. 2018).

This book focuses empirically on offshore *oil* exploration and production off Norway. With increasingly critical and vocal concerns raised about sustainability, climate change, and big oil's dismal record to date of maneuvering in geographic regions with significant social and political challenges, why base this book empirically on oil? As I see it, there are three compelling reasons.

First, in the specific empirical context I draw on, the Norwegian continental shelf with the North Sea, the Norwegian Sea, and the Barents Sea (thus into the Arctic; see figure 1.4), offshore oil and gas are difficult to produce, not to mention find (i.e., oil exploration). The hydrocarbon reservoirs empirically covered in this book reside one to five kilometers below the seabed. They are knowable largely through big, IoT-based data—for example, seismic data (acoustic reflection measurements and processing), well logs (electromagnetic and radioactive measurement of rock properties along an oil well), and production data (measurements of flow volume, temperature, pressure, and chemical composition). The data are inherently uncertain and incomplete. Geoscientists constantly grapple with questions such as: Are the

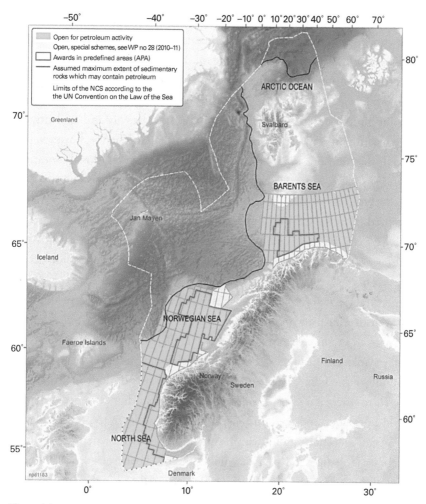

Figure 1.4

Map of the oil licenses on the Norwegian continental shelf.

Source: Reproduced by permission from the Norwegian Petroleum Directorate.

IoT measurements accurate? How do they resonate with other data? What do the data really mean? In a corner of the world where competing globally on the basis of cost (notably wage levels) is prohibitive, oil exploration and production is decisively knowledge based. More to the point, it relies heavily on the knowledge infrastructure of sensor-generated, aggregated, algorithmically manipulated, and visualized big data characterizing all phases of offshore oil—namely, exploration, drilling, and production. In short it provides a vivid case of IoT-constituted, data-driven knowledge work relevant to wide and growing empirical settings and industries significantly beyond the rather limited world of Norwegian offshore oil. In particular, visions and strategies for the future of manufacturing and processing industries, promoted through labels such as Industry 4.0 (Lasi et al. 2014), the fourth industrial revolution (Marr 2016), or industrial IoT,[5] draw heavily and explicitly on digitalization in general and IoT and AI in particular. There is, however, precious little attention paid to the social, technical, and institutional conditions enabling such visions to become more than exactly that—visions.

Second, rather than purging my account of the political controversies vocalized from the early days some fifty years ago and very much ongoing, I include them. Precisely because data-driven "facts" pertaining to oil are never neutral, the machinery of their production and consumption is instructive. The case of Norwegian offshore oil, then, provides a much-needed occasion to demonstrate, not only state, the politics implicated in machineries of knowing. In the institutional and political context covered in this book, national control and regulation, the role of state ownership in oil operators, and, not least, the maintenance of a sustainable commercial fishery in one of the richest fishing grounds on earth have been and very much still are politically contested.

Third, there is a relative paucity of empirical studies from the corporate world in science and technology studies. This is illustrated in the relative absence of any such studies within the MIT Press Infrastructure Series (an exception is Shafiee [2018]). Accordingly, there is, I believe, scope to methodologically and theoretically apply insights generated within the broad church of infrastructure scholars to studies of knowledge production and knowledge infrastructures within the institutional setting of the predominant international business organizations. In an exception to this rule, Bowker (1994) discusses the practices and vocabularies of industry-based scientists and notes a widespread perception that "industry was seen as a second-class choice. It could offer much more money, but at the price of glory and security" (17) This book calls for further attention to what Pollock and Williams (2016) coin a new sociology of how business knowing—so much more than speculation yet less than hard evidence—is made collectively accountable and hence organizationally consequential. To this end, this book draws on my extensive research spanning more than two decades and building on multiple sources of data (see the appendix) from observations, interviews, newsclips (translated from Norwegian by me if not otherwise stated), and documents studies with upstream oil operators (notably NorthOil, a pseudonym), oil service companies, vendors, consultants, and researchers that constitute the industrial cluster or ecosystem around ongoing digitalization.

THE REMAINDER OF THE BOOK

The book is organized into three parts. Part I, "Setting," consists of two chapters that provide a backdrop to the subsequent parts. Chapter 2, "Context," outlines the historic conditions of Norwegian offshore oil. It underscores the formation and later evolution of the institutional fabric of Norway's fifty years of oil exploration. It also details how a small country lacking in experience, knowledge, and capital underwent an unlikely historic transformation whereby offshore oil and gas morphed into a robust industrial ecosystem. A history of viewing the natural resources for energy production (hydroelectric power, oil) as essentially a public good despite opposition and calls for economic liberalization proved crucial to allowing a gradual national engagement with the knowledge infrastructure implicated in offshore oil. The political and institutional processes shaping Norwegian oil and gas have, beyond the general interest, a direct consequence for the main subject matter of this book, digital oil. These political processes resulted in the establishment of "open," noncorporate data sources for, in principle, all data about Norwegian oil, thus significantly shaping the contours of the datafication of digital oil.

Chapter 3, "Apparatus," describes the nuts and bolts of the big data underpinning the machineries of knowing in digital oil. It should be understood as a response to the dangers of overly monolithic, undifferentiated, and generic descriptions of the data that make up "big data." This chapter spells out in some detail the nature and characteristics of big data in oil. The presentation of the many different types of digital oil data is organized by following the different phases of commercial oil activities. Vast, heterogeneous historic as well as real-time sensor data are generated across all phases of oil activities, including exploration (seismic, acoustic reflections), drilling (downhole measurements of pressure and temperature), well logging (electromagnetic, electric, and radioactive measuring into the drilled well), and production (real-time measures of volume, sand detection, temperature, and pressure). More than size or volume (though considerable), the defining characteristic of the data informing knowing digital oil is the notoriously unreliable data from the sensors, the historized and regionalized (silo) nature of the data, the disproportionate emphasis on select "small" data (e.g., slide presentations summarizing an otherwise bewildering situation), and the relative lack of data, resulting in knowledge that is much more than mere speculation yet falls significantly short of hard evidence.

Part II, "Cases," comprises four empirical studies of practices of knowing in digital oil. Again, the sequence of chapters mirrors roughly the phases of commercial oil activities.

Chapter 4, "Data," analyzes the work, or "magic" (Bowker and Star 2000) that goes into massaging, sanitizing, and filtering data to *make* data. Data, despite the imaginary invoked in inflated visions of data science of massive data frictionlessly available at your fingertips (Brynjolfsson and McAfee 2014), is never given but needs to be "cooked" (Kitchin 2014; Gitelman 2013; Jackson 2014; Edwards et al. 2011). Empirically, this chapter focuses on the work of data managers tasked with serving the data needs of geoscientists exploring for oil. Data managers are tasked with the invisible work, which is surprisingly difficult to locate and access in an era in which we all are accustomed to googling, necessary for geoscientists to explore for oil.[6]

Chapter 5, "Uncertainty," zooms in on the work of those at the receiving end of the data managers' efforts discussed in chapter 4. Chapter 5 analyzes the work of geoscientists exploring for oil. The truly knowledge-intensive part of deepwater offshore oil is locating it. Having positively located it, developing and producing it is, relatively speaking, less challenging.[7] Oil exploration is radically underdetermined by data. Data are coarse (seismic), scarce (well data), and inherently uncertain. Exploring for oil is all about pragmatically making the most of what you have. This chapter describes the trajectory of a *prospect*—a candidate for an oil reservoir—through a process of gradually accumulating evidence (credibility, trust) that sometimes get undercut by (sufficient to make a difference) inconsistencies in the data. Never gravitating toward closure or consensus, there stubbornly remains a multiplicity of diverging interpretations of the prospect—a multiplicity that fills generative and hence productive roles.[8]

Following the phase of oil and gas exploration covered in the preceding two chapters, a selected few fields are developed and subsequently put into production. Chapter 6, "Knowing," analyzes one aspect of production. It addresses a potentially devastating problem that occurs during everyday oil production: sand. With most of the North Sea hydrocarbons trapped in Jurassic sandstone, the potential of sand to be part of the produced flow of hydrocarbons is immanent. Sand erodes pipes, valves, and chokes and, left unchecked, represents grave risks. The focus in this chapter is on the successive stages of the digitalization of sand-monitoring work routines: sand that is physically present in inspection cups attached underneath the production pipeline at selected sites is subsequently replaced by digital renderings of sand, providing sound or electroresistance sensor readings, plotted trends of measurements, and a predictive simulation model. *What* we know about sand is invariably tied up with *how* we know it. Digital sand "stands in for" physical sand—that is, sand is as much physical as digital from the point of view of knowing or acting upon sand (cf. Latour's speed bumps).[9]

In chapter 7, "Politics," the politically charged nature of the machineries of knowing digital oil is explicitly addressed. Oil activities, despite their socioeconomic significance in sustaining living standards in a high-cost-of-living country within a publicly financed welfare state,[10] are and have been highly controversial throughout their fifty years of history in Norway. Nowhere is this more apparent than the still open controversy over whether to lift the

Table 1.1

A road map for how and where the three aspects of knowing digital oil (objects, modes, and machineries of knowing) are addressed in part II of the book, "Cases." Dark gray indicates the main focus, while light gray indicates a supplementary but not main theme.

	Objects of knowing: data	Modes of knowing: algorithmic and data driven	Machineries of knowing: infrastructure and platforms
Chapter 4: Data	■ (dark gray)		▨ (light gray)
Chapter 5: Uncertainty	■ (dark gray)	■ (dark gray)	■ (dark gray)
Chapter 6: Knowing	■ (dark gray)	■ (dark gray)	■ (dark gray)
Chapter 7: Politics	■ (dark gray)	■ (dark gray)	■ (dark gray)

present ban on oil activities in the Arctic in the areas of Lofoten, Vesterålen, and Senja, as well as in the high north of the Barents Sea, which is abundantly rich in fish. The chapter traces the emergence of NorthOil's efforts to know the marine environment through IoT-instrumented installations in an attempt to preemptively counter the criticism of oil activities in these areas. Crucially, this chapter details the negotiated, constructed, and evolving nature of the IoT-generated machineries for producing facts about the marine environment.[11]

Given the characterization of the three aspects of knowing in digital oil summarized above—objects, modes, and machineries of knowing—table 1.1 provides a road map of the empirical chapters in part II.

Part III of the book, "Implications," consists of chapter 8, "Conclusion," which synthesizes a perspective on digitalization as efforts to quantify the qualitative. In this chapter I elaborate and spell out the implications of such a perspective along the three dimensions above: quantifying the objects, the modes, and the machineries of knowing.

The appendix, "A Note on Method," offers some reflections on methods. Leaving the details of data collection to the scientific research literature, this chapter instead discusses the longitudinal nature of my research, together with the ambition to comprehend the full ecosystem of digital oil, rather than singular organizations.

I SETTING

2 CONTEXT

This chapter provides a historic outline of the political and institutional processes that shaped, and still shape, the organization of Norway's oil and gas industry. There are compelling reasons why this context is significant and relevant to my analysis of datafication in oil and hence provides the motivation for this historic outline.

First, the (evolving) configuration of the industrial ecosystem is difficult or impossible to grasp without such a context. More specifically, policies, institutions, and regulations have actively promoted a nationalist industrial agenda. This is the key reason why the oil industry in Norway has only slowly been woven into the once significant maritime industrial cluster: it has been intentionally slow to allow local learning to take place to avoid oil in Norway largely being run as an outsourced activity of international, notably US, global companies. This nationalist industrial agenda, then, established the conditions for what over a few decades evolved into a robust, comprehensive, and heterogeneous industrial ecosystem around Norwegian offshore oil. The extensive Norwegian-based industrial ecosystem presently in oil is the cumulative result of sustained politically and institutionally imposed conditions for locally growing machineries of knowing oil.

Second, a formative feature of Norway's policies on offshore oil is how they were approached through the lens of historic policies on the governance of energy (hydroelectric) and natural resources (minerals): oil resources were governed by recognizing them as *public goods*, not private property. Consequently, the datafication of oil—the vast and accumulated data sets generated from oil and gas exploration and production (see details in chapter 3)—had to be made openly and publicly available, not kept as private corporate assets. These policies were part of a political push to promote collaboration across different oil companies and to lower the barriers to entry to new ones.

Third, the political emphasis on moving slowly enough for Norwegian-based actors to learn the necessary know-how, skills, and practices from their international partners implied that the maritime cluster of small workshops, vendors, industries, service providers, research institutions, and certification bodies were able to get onboard. Building on a rich history in the maritime industries, this resulted in Norway taking a global lead specifically in offshore *subsea technologies*—that is, unmanned, remotely operated facilities (see figures 1.2 and 1.3). Significant to my analysis of knowing digital oil is the pivotal role of sensor and Internet of Things (IoT) data, the only type of data available from unmanned subsea operations.

There are competing ways to present the historical background of oil and gas in Norway. One account would emphasize the elements of serendipity. Hydrocarbon reserves result from the coming together of a number of geological conditions. In the North Sea, the formative conditions began during the geological era of Jura some 150–200 million years ago. The North Sea was a big shallow basin that allowed large quantities of organic matter in the form of phytoplankton to fall to the sea floor after death, a necessary but nowhere sufficient condition for oil and gas to form. Subsequent elaborate structural and local geological processes—sedimentation from rivers gradually sealing the organic matter into rocks compacted over time and pushed deep, but not too deep, into the earth as oil forms at 60–120 degrees Celsius in the *source rock* due to the thermogenic breakdown (cracking) of organic matter (kerogen); a *migration path* through sufficiently permeable rocks or fault lines resulting from the movement of tectonic plates to be transported and, crucially, avoiding the fate of most hydrocarbon, which is to leak out or evaporate; and a *trap* in the form of nonporous rock blocking and gradually accumulating hydrocarbon—are, globally, relatively rare (see figure 2.1). In short, you need the geological conditions of what in the oil and gas vocabulary is known as the hydrocarbon "kitchen" (Schlumberger 1998).

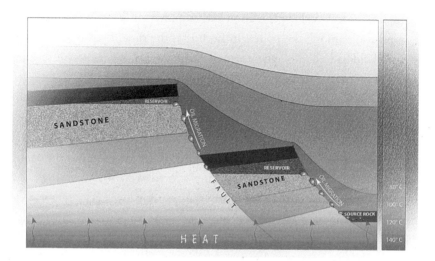

Figure 2.1

A geological trap with the three necessary conditions of a source rock, a migration path, and a seal.
Source: Reproduced by permission from the Norwegian Petroleum Directorate.

In the North Sea, a gas discovery by Shell and ExxonMobil in 1959 in the Dutch sector triggered interest in oil exploration. If present in the southern corner of the North Sea, could it then not also be present further northwest into the British sector, further northeast into the Danish sector, or, the empirical focus of this book, further north into the Norwegian sector?

In contrast to focusing on geological idiosyncrasies, an alternative way of understanding the history of oil and gas in Norway underscores the historic legacy in Norwegian political economy around the perception and governance of natural resources.

Norway has a long and rich history of mining for minerals. For instance, the silver mine in Kongsberg is from 1623, and the copper mine at Røros is from 1645. To enable hydroelectric installations in the mountains, Norway was one of the first countries in the world to establish, in 1858, a national geological survey agency, Norges Geologiske Undersøkelser (NGU). The explicit agenda of the NGU, part of and in preparation for the nationalism behind the independence from Sweden some fifty years later, was to contribute to the industrialization and modernization of Norway by identifying natural resources. Value generation and employment were in dire need for a population among the poorer in Europe at the time. During the 1800s, only Ireland had a higher relative emigration than Norway of people seeking brighter economic prospects in America.

However, tasked in 1958 with evaluating the chances of finding oil within the Norwegian sector of the North Sea, the NGU was unmistakably dismissive. In a letter to the Ministry of Foreign affairs, it concluded that "you may disregard the possibilities for the presence of coal, oil or sulphur on the continental shelf along the Norwegian coastline" (Carstens 2014). Accordingly, the enthusiasm and expectations for Norwegian oil were low. Lacking in experience, technology, and capital, no Norwegian companies were willing to risk investing in oil exploration in the Norwegian part of the North Sea. Among the world's earliest global industries, the American oil operator Phillips Petroleum in 1962 approached the Norwegian government with an offer to start exploring for oil. Proposing to pay USD $160,000 per month, the offer was understood as an attempt to gain exclusive rights. The offer was turned down.

There are, undeniably, considerable elements of chance and idiosyncrasies in any history of oil in Norway. Nevertheless, an overemphasis on chance is strongly misleading. It leaves a number of questions unaccounted for. Numerous countries in the world enjoy at best a mixed blessing, giving rise to the notion of the "oil curse," a period of economic fortune that is only temporary. One of the handful of oil bureaucrats involved in the establishment of the Norwegian Petroleum Directorate (NPD) pointed out: "Poor countries dream of finding oil like poor people fantasise about winning the lottery. But the dream often turns into a nightmare as new oil exporters realise that their treasure brings more trouble than help" (Sandbu 2009). As critical studies compellingly document, oil has in many if not most cases fed corruption, inequality, and cronyism. Petrostates, especially after the 1973 oil crises, note Reyna and Behrends (2011), "are capital-intensive oil exporters with high ratios of oil to total exports; petroleum industry enclaves; and enormous rents or royalties (from oil sales), which accrue directly to the central government. . . . Oil turned out to be a development 'curse'" (5).

Also, so-called developed countries with relatively mature institutions and governments regularly confront dramatic declines in manufacturing capacity in nonoil industries resulting from the influx of income from natural resource (oil) exploitation. The "Dutch illness," a term coined for the situation prevailing in the Netherlands following the North Sea gas discovery, may easily emerge given the temporary and uneven but substantial injection of income from oil into national economies. This wreaks havoc with the economy, with staggering inflation and long-term loss of productivity. Current candidates include Mexico, Venezuela, and Brazil.

How was it that Norway was able to maintain national control, unlike its otherwise-close Scandinavian neighbor Denmark? Accepting an offer from A. P. Møller similar to what Phillips Petroleum had offered Norway, Denmark granted exclusive rights to the oil company. How did a present-day, internationally competitive industrial cluster within the whole ecosystem around oil—oil operators, technology vendors, maintenance, financial institutions, engineering, insurance, and service providers, including software-based ones—emerge from a situation in the late 1960s in which neither Norwegian industry nor national authorities and institutions had the slightest experience or knowledge about what was involved in oil? As has attracted the attention of international media (Holter and Sleive 2017), how has Norway established a national sovereign trust fund worth USD $1 trillion and controlling about 1.3 percent of the world's traded equities while also flexing its muscles to pressure deinvestments of polluting industries such as coal and, not without irony, debating whether to also deinvest in oil and gas (Bloom 2019)?

Falling significantly short of a substantial history (for a fuller account, see Ryggvik 2009, 2015), what follows traces the historic conditions behind the establishment and evolution of the institutional fabric, the interplay between national authorities and private businesses that underpins Norwegian oil and is helpful in understanding the chapters on practices of knowing in digital oil in part II of this book. The outline provides an institutional perspective on a history that, perfectly analogous to the geological conditions indicated above, involved the coming together of a wide set of historically contingent conditions. Given its widespread perception of being successful, there is significant interest in understanding, not to mention mimicking, the Norwegian history. Lula, then president of Brazil, for instance, visited

Norway in 2008 with the explicit goal of learning from its institutional history in order to mimic it in Brazil (E24 2004).

As economic historian Einar Lie (2017) notes, however, imitating the institutional history of Norway during its fifty or so years of involvement in oil is rather difficult given its highly localized, historized conditions—conditions leading Lie to conclude that "nobody can replicate the Norwegian success in oil. Not even we [Norwegians]."

In the context of this book, the historic outline serves three purposes that operationalize the initial motivations cited at the beginning of the chapter. First, it unpacks the variety of actors (oil operators, oil service providers, consultancies, public agencies, research institutions) that make up "the" industry of Norwegian offshore oil. Second, it sheds light on issues of *data friction* that, despite policies of open oil data, make the accessing, processing, and sensemaking of oil data challenging. Third, the prominence of the unmanned subsea production characteristic of Norway's offshore oil shifts sensor and IoT-based oil data to the forefront of the analysis.

NATURAL RESOURCES FOR ENERGY PRODUCTION AS A PUBLIC GOOD

The political inclination not to privatize the oil fields, as Denmark did, runs deep in Norway. Formative experiences included regulation of the previous natural resource-based energies, notably hydroelectric power. In order to ensure the collective public benefit of hydroelectric power plants, they are by specially tailored legislation leased to private developers but eventually must be returned to public ownership (*hjemfallsretten*) after fifty to seventy years. This legislation granting private companies temporary control over hydroelectric plants was, not without considerable controversy, passed by Parliament in 1909 (Norwegian Ministry of Oil and Energy 2004). Oil and gas, too, were approached through this lens of a public good. The licenses granted for oil and gas as regulated by petroleum legislation and exercised through the concessional rounds were granted for only six years, with the additional demand that one-quarter of the license be handed back after three years.

This facilitated a phased entry of Norwegian companies, after international companies initially were allocated the licenses.

The perception of natural resources as a public good was broadly shared across the political spectrum. Phillips Petroleum's offer of oil exploration in 1962 was turned down because the then Labour government insisted that a legislative framework for oil and gas had to be in place to regulate the role of private companies in the spirit, if not the letter, of the regulation of hydroelectric power production. This heritage is still apparent. There is also an ongoing debate whether fish farming, Norway's second-largest source of export revenues, should be regulated as a public good like hydroelectric power and oil. The argument, which is exactly the same as that regarding hydroelectric and fossil energy, is that fish farming generates private revenue based on what is essentially a public good, the sea (Bjørnestad 2019).

Mimicking the US decree ten years earlier declaring federal jurisdiction of the outer continental shelf, the Norwegian government in 1963, seven months after the offer by Phillips Petroleum, declared sovereignty over the Norwegian continental shelf. A crucial organizing vehicle to regulate oil activities was the allocation of licenses after application, unlike in the US, through market-based auctions. This system is known as *concessional rounds*. By exercising control over the concessional rounds, Norwegian authorities laid out duties and incentives (including a favorable tax regime) for oil companies. In the first concessional round in 1965, seventy-nine blocks (a block comprises an area of about five hundred square kilometers) were allocated, and practically all international oil companies took part. The first wells drilled were dry or had reserves that were not commercially viable. Enthusiasm dwindled. Phillips Petroleum's oil find, Ekofisk, in 1969 effectively marks the turning point and the beginning of the oil industry in Norway.

Keenly aware of the fact that Norwegian industry from the outset lacked the experience and know-how to take a prominent part in the oil industry, political strategists shaping Norwegian industrial policies aimed for phasing in Norwegian industrial involvement *gradually*. The Parliament white paper (Stortingsmelding 1974) in 1974 made clear that "given our concern for long-term resource exploitation and a broad assessment of the societal consequences the Government concludes that Norway should adopt a moderate tempo in the extraction of the petroleum resources" (25). As evolutionary economist Reinert (2007) points out, Norway did what most rich countries have done historically—namely, regulate the full force of free-market economics until Norway was able to compete on an even pitch:

> The great debate over economic policy at the time [in the US and UK in the 1850s] was not whether one should protect industry—almost everybody could agree to that—but how this should be done. Today the frail industry of the Third World is being suffocated by the same free trade Norway was able to defend itself against for a century. The fact that Norway is in need of free trade *today* neither means that it needed it 150 years ago, nor that poor countries need it now. (59)

At the core of Norwegian policies, then, was the recognition that Norwegian participation in oil had to be understood as a catch-up—that is, a *learning* process. Learning, with important elements of learning by doing, would necessarily take time. Accordingly, a central aspect of Norwegian oil policies, inscribed into the 1974 Petroleum Development Act passed in Parliament, was the "moderate tempo" of Norwegian oil activities. Against the push for the full-throttled endeavor advocated by international oil companies, Norwegian authorities, through the concessional rounds, actively regulated the pace of the oil industry to allow Norwegian companies time to learn the ropes of the trade. The maritime industries were especially important. The maritime industrial cluster had historically demonstrated, in many ways, a surprising ability to transform radically via several technological paradigms over the last couple of centuries. Apparently, at every turn all businesses in the maritime cluster went bankrupt and the employees dispersed: a cluster on the south coast tied to sail and wooden hulls in the late 1800s withered and died before a new cluster in eastern Norway tied to steel hulls in the mid-1900s emerged, which, yet again, evaporated when a new industrial cluster in northwestern Norway tied to catamaran hulls emerged. In such a history of apparent discontinuities, what accounts for this continuity is a question posed by technology historian Andersen (1997). The continuity, he forcefully argues, was the then still operating *knowledge infrastructure* consisting of a set of institutions that included banks, insurance companies, shipping brokers, research institutions, and certification companies.

The "moderate" pace of Norwegian oil activities during the first couple of decades of oil in Norway allowed Norwegian shipyards, rig owners, supply vessel owners, maritime engineers, and maintenance/service providers to morph into what by the beginning of the 2000s was a significant oil ecosystem. In 2017 there were about 140,000 employed in the sector, accounting for 14 percent of gross national product (GNP) and 37 percent of the country's net export value (see table 2.1; Norsk Petroleum 2021). Robust innovation clusters, especially around drilling in the south and on the west coast, emerged. Illustrating the learning approach, Norwegian activities were initially geared toward construction of the rigs and the equipment required for oil and gas activities, not the core activities of exploring for or producing oil.

Since the mid-1990s there has been a sustained effort to enact industry-wide standards to promote efficiency. Since 1995, Norwegian subcontractors have committed to deliver, according to NORSOK standards (Standards Norway 2021), a set of standards for specifying and providing equipment to the oil and gas industry. There are presently close to one hundred standards. However, as a broad consortium of industry actors documented in 2018, there is still a long way to go to break out of organizationally defined, digital silos (Konkraft 2018). Alongside standardization efforts targeting requirement specifications are efforts to standardize the vocabulary and specification of the subsurface itself with the Open Subsurface Data Universe (OSDU), by far the most forceful with a significant and expanding set of industry members (OSDU 2021). These standardization efforts underpin

Table 2.1

Summary of the economic significance of oil and gas in Norway along with the number of people employed in the oil and gas industry in Norway. *Source:* SSB.no and Norsk Petroleum.

Economic indicator	Figures from 2017
GNP	14 percent of GNP
Net export value	USD $45 billion, about 37 percent of total export value.
Number of people employed	139,500, or 5.1 percent of total employment.
Investments in oil	72 percent of investments go to domestic service providers and vendors. Investment in oil is about 19 percent of total industrial investments.

the industry-wide transformation of the whole industrial ecosystem, fueled by digital platforms embedding these standards (Konkraft 2018).

The Journey from the Gulf of Mexico to the North Sea

The transfer of ideas, practices, or technology from one cultural context to another is anything but straightforward. Latour (1987), in a critique of the mechanistic and deterministic assumptions underpinning the notion of "diffusion," argues that it is more about translations. Similarly, taking the improbable travel of the game of cricket from stiff-upper-lip, Etonian England to a game played on every vacant lot or street corner in India as an example, Appadurai (1996) discusses the processes whereby "transfer" is achieved:

> The indigenization of a sport like cricket has many dimensions. It has something to do with the way the sport is managed, patronized, and publicized; it has something to do with the class background of India players and thus with their capability to mimic Victorian elite values; it has something to do with the dialectic between team spirit and national sentiment, which is inherent in the sport and is implicitly corrosive of the bonds of empire; it has something to do with the way in which a reservoir of talent is created and nurtured outside the urban elites, so that the sport can become internally self-sustaining; it has something to do with the ways in which media and language help to unyoke cricket from its Englishness; and it has something to do with the construction of a postcolonial male spectatorship that can charge cricket with the power of bodily competition and virile nationalism. Each of these processes interacted with one another to indigenize cricket in India. (90–91)

In many ways the transfer of oil practices from the Gulf of Mexico to the Norwegian continental shelf was as improbable as that of cricket. Oil exploration and production, an initially onshore endeavor, had by the 1980s shifted into the shallow (twenty to fifty meters in depth) waters off Louisiana and Texas. Dominated by work organizations in the southern US unaccustomed, if not outright hostile, to the cornerstone of Norwegian/Nordic work-life traditions made up of unionized, regulated, and negotiated agreements worked out between employers and employees, the travel was not unlike Appadurai's cricket. Initially completely dominated by American companies and work-life traditions, Norwegian companies and employees, as a result of the allocations in the concessional rounds, gained a steadily increasing footprint. Starting as minority partners, Norwegian companies over a period of a couple of decades acquired more central roles. As pointed out, the dominant political sentiment in these formative years of Norwegian oil in the 1970s was toward a highly regulated rather than free-market regime. As Cumbers (2012) observes when comparing the UK experience with oil to the Norwegian one, "Norwegian trade unions remain important actors (beyond the wildest dreams of their UK counterparts) both in the political sphere, through their influence on the ruling Labour coalition, and in the economic sphere, through their role as social partners" (238). This should not be misconstrued as a consensus. The shipowners' organization, a significant political influence in Norway given its long history in shipping, systematically lobbied for privatization, the increased presence of private companies, and free-market regulation.

The legislative foundation of Norwegian oil worked out in the 1970s was strongly influenced by the nationalism invigorated as part of the 1972 referendum that decided to keep Norway out of the European Union (EU, or EF, as it was called in Norway at the time). Pitched very much as an issue about national sovereignty, the 1972 referendum boosted national control. The petroleum legislation from 1974 thus made clear that all companies needed to comply with Norwegian work-life regulation. Even with a conservative prime minister in office, the international oil operators were told in uncharacteristically clear language that they were expected to join the employers' organization, which automatically put them inside the institutional negotiating framework with the unions.

The relatively forceful regulation imposed by Norwegian authorities met with considerable opposition from the international oil companies. How did a small country such as Norway enforce its will and policies? The oil crisis in 1973 proved vital. It significantly shifted power from oil operators to oil-producing countries. Norway seized the opportunity to tighten its grip on the oil industry, including hiking up taxes. Without the new political situation emerging with the oil crises, Norway's negotiating power relative to the international oil companies would have been significantly weaker.

Several Norwegian oil companies were established in the 1970s to 1990s, the majority of which were private (Hydro, Fred. Olsen, and Saga). One of these, Statoil (later renamed Equinor), was governmentally owned when established in 1972. Consistent with the strategy to learn the trade over time, Statoil first targeted downstream (transportation pipelines, processing, refinement, retail) rather than the more knowledge-intensive upstream activities (exploration, field development, drilling, production). The first oil field Statoil itself found rather than merely being a co-operator in an allocated license was Norne, which was discovered in 1992.

In parallel with Statoil/Equinor, the Norwegian petroleum authorities established the NPD. This institution was tasked with evaluating the bids for the different concessional rounds. To do so, they required all oil companies to share their current knowledge about prospective oil finds as part of their bid. This de facto allowed the NPD to generate a cumulative overview of the whole continental shelf. In a move starkly different from the privatized data on US-based oil, the data about oil in Norway, cumulatively and collectively gathered by the companies, was made openly available as a common good (NPD 2021). This was intended to spur further collaboration and trigger the identification of new opportunities. Similarly, it institutionalized an obligatory archive of core samples that had been extracted from drilling operations since 1965. When needed, it could and would also order its own seismic surveys on demand to supplement the commercial ones (Ryggvik 2009).

ECONOMIC LIBERALIZATION

Organized around the overriding concern of maintaining national control of natural resources, the political atmosphere changed markedly during the late 1980s and early 1990s. Strong international trends toward liberalization prevailed, and Norway was no exception. Norway, not a member of the EU, had negotiated a trade agreement with the EU known as the European Economic Area (Europeiske Økonomiske Samarbeidsområde), which called for the lifting of protectionist arrangements that until then had shaped Norway's regulation of oil and gas. In a fundamentally different position by the early 1990s than some two decades earlier, Norwegian companies were—finally—ready to face international competition. As rich countries have made their habit, liberalization and free-trade policies are advocated only after a period of learning protected from these same policies. England, honing its expertise in steam engine design, lifted its protectionist regulation and pushed for a "free" market only after securing a healthy head start. At the core of dominant economical thinking, Reinert (2007) argues, there is a gap between, on the one hand, the rhetoric of free trade and competitive advantage economics and, on the other hand, the policies employed in practice: "In practical terms, then, lofty economic rhetoric is for export to others, while completely different pragmatic principles are adhered to for the realities at home" (22–25). Reinert continues, "Don't do as the Americans tell you to do, do as the Americans did," as exemplified when "following England's practice rather than her theory [of free trade], the United States protected their manufacturing industry for close to 150 years."

In the aftermath of the "roaring" 1980s, with its staggering inflation and struggling banks, the Norwegian Parliament finally agreed in 1990 to establish a national sovereign wealth fund. Motivated by the hard-won economic lessons from the 1980s, the near consensus in Parliament was for the fund to invest only abroad in equity and property. In a direct response to avoid the Dutch disease, the fund's revenues would be buffered from the national economy by a vehicle known as the *handlingsregel*, which literally means the "rule for action." This rule specifies that a maximum of 4 percent of the *revenues* of the fund's investment are available for national budgets. With the value of the fund at USD $1 trillion, this rate has recently been reduced to 3 percent.

An important reason why Norwegian oil companies were competitive with the unleashing of liberalization during this period was the number of technological innovations, developed during a more protectionist period, addressing the complexity of offshore oil on the Norwegian continental shelf. During the 1980s and 1990s, the North Sea was the most challenging offshore operation in the world. Offshore oil in the Gulf of Mexico at the time was in shallow waters. Technology, practices, and knowledge of these shallow waters largely replicated its onshore, historic origin. With the Norwegian continental shelf at more than one hundred meters deep and, more importantly, with oil reservoirs more than one to five kilometers below the seabed, the limits of offshore oil were pushed. Increased depth implies an increased complexity of operation. In this period the Norwegian continental shelf was effectively a real-world innovation laboratory testing the limits of increasingly complex deepwater offshore oil operations (Ryggvik 2015). Thus, its harsh weather conditions and challenging offshore settings proved to be a comparative advantage for Norwegian industry. Innovations in multiphase meters (allowing the transportation of hydrocarbon from greater depths), water injection (upholding reservoir pressure to prolong production), horizontal (rather than vertical) drilling to bring down drilling costs, and, not least, the pioneering emphasis on subsea—unmanned, IoT controlled—production facilities residing on the seabed gave the Norwegian industrial oil ecosystem a competitive position. Consequently, Norway's offshore oil—the empirical setting of this book—has sensor and IoT data at the core.

Peaking in 2001 at 3.4 million barrels of oil, making Norway one of the largest oil producers in the world at that time, production has since fallen. This triggered the establishment of a new regime in 2003 in "mature" areas (*forhåndsdefinerte områder*). Encouraging smaller operators and innovative technologies stimulates a lightweight approach. Exploring areas close to existing transportation infrastructure, for instance, is encouraged, as this significantly pushes the cost of production down by a factor.[1] Simultaneously, incentives to promote oil exploration have been introduced, as well as a tax deduction scheme that lets the oil operators have 78 percent of costs deducted.

SUBSEA TECHNOLOGIES: A COMPETITIVE ADVANTAGE?

The emphasis in Norway on subsea, IoT-based, remotely operated oil production is central to this book. It is a defining aspect of the big data connotation of digitally rendered oil. In subsequent chapters, a set of selected knowledge-based work practices are analyzed to uncover how they are shaped—indeed, to a large extent constituted by—the machineries of digitally rendered "reality." It is not a distant vision. By the early 2000s, most production wells on the Norwegian continental shelf were tied to subsea installations. One of the smaller private oil operators, Aker BP, is actively pursuing an agenda of a fully digital, remotely operated oil installation in one of their smaller fields, Ivar Aasen (Aker BP 2021).

Oil companies are pushing for increasingly digitally enabled, if not outright automated, operations. Extending the scope of IoT-based monitoring and control is central to these visions. The visions, however, meet with considerable opposition. The unions, concerned about future employment, argue that the shift toward IoT-based, unmanned, automated operations represents a threat to safety (see quote in chapter 1). They point out that with the phasing out of the initial American domination in Norwegian oil, safety indicators (number of accidents, deaths) are lower in the Norwegian institutional setting than elsewhere. Safety issues plagued the nascent oil industry in Norway, especially during the first couple of decades. From 1965 to 1978, 82 workers died. In 1980, with the disastrous Alexander L. Kielland accident, 123 died. With increasing Norwegian involvement in all phases of offshore oil, the death toll has come down.

The Obama administration's National Commission on the BP *Deepwater Horizon* Oil Spill and Offshore Drilling (2011), established to investigate the Deepwater Horizon blowout in the Gulf of Mexico in 2010, noted this decline with interest. In a discussion of the similarities and differences in the institutional regulation of safety between the Norwegian continental shelf and the Gulf of Mexico, the members wrote: "From 2004 to 2009, fatalities in the offshore oil and gas industry were more than four times higher per person-hours worked in U.S. waters than in European waters, even though many of the same companies work in both venues" (225). However, Ryggvik (2015) warns against jumping to conclusions. Significant differences between the Gulf of Mexico and the Norwegian continental shelf make a direct comparison of accident rates problematic. In addition, in 2010 a serious incident triggered an audit by the Norwegian Petroleum Safety Authority (Petroleumstilsynet) concluding that "marginally different circumstances" would have resulted in a blowout with a damage potential at least on the same scale as the Deepwater Horizon disaster (Nilsen and Stensvold 2010). So perhaps luck as much as institutional regulation accounts for better safety statistics?

Safety concerns in a Norwegian setting arise not only as outlined above between oil companies, national authorities, and the unions. Worries about safety, very much including concerns for the environment and climate change, have been a central dividing line in Norwegian politics throughout its fifty or so years of history with oil. They are closely tied to concerns for threats to Norway's commercial fishing, the country's second-largest source of export revenues. The question of where, if at all, to allow oil and gas activities was and still is controversial. The Conservative and Liberal Parties, currently and earlier in Norway's history with oil, are the most enthusiastic proponents of oil and gas. The left and center parties are largely skeptical, typically underscoring the environmental hazards and climate change concerns. The Labour Party, traditionally a dominant influence in Parliament, is split between a camp emphasizing industrial value generation and employment versus a camp sympathizing with environmental concerns. A principal means of regulating, subject to shifting majorities and alliances in Parliament, is through the concessional rounds. Until 1979, oil and gas activities north of the 62nd parallel were banned. Parts of the North Sea and the Barents Sea were opened later in 1993 and 2007, respectively, with an expanding set of blocks allocated following that. Key areas in the North Sea (Lofoten, Vesterålen, and Senja) as well as in the Barents Sea are—for the time being—off limits to oil and gas activities.

CONCLUSION

The historic outline of the political, institutional, and industrial processes during Norway's roughly fifty years of offshore oil and gas has significance and implications for the context and content of knowing digital oil. It provides insights into the institutional fabric of those involved, what data underpin it, where these data come from, and how sensor data take on such a prominent role. To flesh this out, chapter 3 details the types of oil data involved and, crucially, the issues involved in grappling with oil data; the subsequent chapter, against the outline provided here, unearths the socially, institutionally, and technically imbricated data of digital oil.

3 APPARATUS

The fundamental challenge of developing a geological understanding of a particular area is to reconstruct a process spanning geological time frames (millions if not billions of years) from your collected data or evidence on the current situation. You *infer* a process from presently available data or, metaphorically, a process from its end product. Developing a geological understanding thus takes the form of a process of provisional and fallible sensemaking. It is also deeply qualitative because the way it has traditionally been taught underscores the importance of tactile and visual experiences of geological phenomena. Field trips are central to practicing these skills. Early tools such as a hammer, a magnifying glass, handmade sketches, and classification templates of rock coloring, layering, and porosity pioneered in onshore settings help cultivate these qualitative skills (Frodeman 1995, Almklov and Hepsø 2011, Latour 1999).

The phenomena under study in this book, offshore hydrocarbon reservoirs off Norway, are not available for geological field trips. Residing kilometers below the seabed at several hundred meters deep, they are, for all practical purposes, neither directly perceivable nor accessible. They are largely known through a variety of sensor-based, instrumented measurements, thus highlighting Bowker's (1994, 4) concern that "few historians have looked at what is entailed in producing a measurement—in simultaneously defining a unit of measure and doing all the infrastructural work that is involved in maintaining that unit."

The everyday practices in present-day commercial oil activities offer rich opportunities to trace the interplay between the somewhat caricatured qualitative sentiments and the quantification inherent in digital representations of geodata. Rather than the one-way street of quantification of the qualitative, knowing in digital oil implies resurfacing and remaking the qualitative within the quantitative. As Monteiro et al. (2018) observes, commercially based exploration geologists, despite an abundance of digital tools and data, are quick to resort to drawings and sketches (cf. Ihde 1999). Although their everyday practices are dominated by grappling with digital representations of geodata, field trips to a geological *analogue*—that is, a formation similar to the area you are primarily interested in—are still organized. For instance, a geological analogy to parts of the Norwegian Sea is, after the tectonic plates shifted, on the east slope of a mountain in present-day Greenland (Almklov and Hepsø 2011). Similarly, figure 3.1 depicts a geologist sketching from an analogue during a project I was involved with in the buildup to this book.

Figure 3.1
Hand-drawn sketches illustrate nondigital tools in geology. Picture is taken from a visit to a *geological analogy*, an accessible location somewhere else in the world similar in important ways to the area you are principally interested in.
Source: Photo by Irina Pene.

Still, to an increasing degree,[1] geological understanding of a phenomenon of interest, the geological formation, is predominately grounded in the digital geodata. Knowing and working with upstream commercial oil operations are largely knowable through a comprehensive machinery (infrastructure) of digital tools, practices, and data; in short, knowing is increasingly about *digital* oil. Given that, it is vital to get into the details of the data that make up geodata.

DATA TYPES AND CHARACTERISTICS

Geodata are anything but monolithic. Data come in a variety of *types*; are generated at different *phases* of oil activities using a variety of *measurement instruments* (sensors); are captured for various *purposes*; are produced by *organizations* with varying degrees of *real-time* (stream) quality with dramatically different data *quality* (uncertainty or noise); are of considerable *size*; and, once captured, are algorithmically manipulated and visualized in numerous *information systems*. Data, in short, have distinctive Internet of Things–based, big-data qualities vital to grasping *how* knowing digital oil is done; oil is thoroughly datafied (Lycett 2013; Hoeyer et al. 2019).

The purpose of this chapter is to provide the necessary background to what the machineries of knowing, detailed in later chapters, draw on in their data-driven practices. To get at knowing in digital oil, it is crucial to appreciate the considerable extent and variety of infrastructural work—sanitizing, assuring quality, triangulating, synthetically generating missing data, and deliberating peers—involved.

To provide an overview of the rich variety and characteristics of the big data constituting digital oil, it is helpful to consider the phases of commercially based upstream oil and gas. For this purpose, I focus on the following broad phases (see table 3.1 for an overview of corresponding digital tools):[2]

- exploration
- drilling
- well logging
- production
- monitoring, maintenance (and later abandonment)

Table 3.1

Summary of key digital tools for different geoscience disciplines.

Type of tool	Number of systems	Example of functionality
Seismic processing	3	Velocity determination and model building
Seismic interpretation	3	3D viewing and interpretation to support fault and horizon interpretations
Petrophysical evaluation	1	Load and manage well logs
Geological modeling	4	Correlate well logs, build cross sections
Drilling	2	Well planning and monitoring
Reservoir technology	1	Reservoir simulator
Production technology	5	Network of surface gathering from field production systems
Specialists' tools	11	Trap testing (interpret, model, and validate traps)

Exploration   Reflection seismology (a seismic survey) is a method to estimate the properties of the subsurface up to several kilometers below the surface of the earth or, in the case of offshore oil, the seabed. It operates logically in the same way as sonar or radar. It requires a seismic source, such as an air gun, that generates a signal strong enough to penetrate significant subsurface depths and detectors to register the returning reflections in order to construct the seismic record. Traditional offshore seismic surveys were shot with vessels towing one or more cables, known as streamers. Two-dimensional (2D) seismic surveys use one streamer (see figure 3.2), while 3D surveys use up to twelve streamers, reconstructing three dimensions by combining and triangulating between the individual streamers. A streamer is several kilometers long, making seismic survey vessels the largest man-made moving objects on the planet. Seismic data are voluminous. Seismic data in 3D were first used on the Norwegian continental shelf in the 1980s. Seismic data are digital recordings of the reflections of acoustic waves sent down to the reservoir in seismic surveys. Via specific arrangements of sound sources and microphones and extensive computer processing, geophysicists can outline layers and other structures in the rock based on contrasting acoustic properties. Most importantly, such contrasts reveal density differences

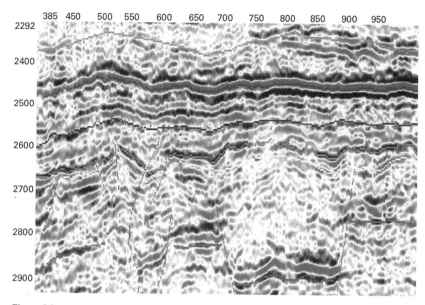

Figure 3.2
Seismic horizons from the Sleipner field.
Source: Reproduced by permission from the Volve open data set.

between porous and nonporous rock, the former being more likely to contain hydrocarbons.

Seismic data (*prestack*) are filtered and processed as part of the labor-intensive work of seismic interpretation. Typically, less than 1 percent of the source data from a seismic survey are used (to generate *poststack* seismic data). The rest are considered noise or noninterpretable wave phenomena filtered out by nonlinear mathematical methods. Large data sets are generated by seismic surveys. One survey yields about one terabyte. There are significant time delays involved with seismic data. A seismic survey campaign generating new data lasts for days. The data then have to be filtered, washed, and analyzed by the service company running the seismic campaign before they are handed off to the client, the oil operator(s). Seismic services and their subsequent processing are examples of the extensive outsourcing that international upstream oil operators employ, giving rise to a thriving ecosystem of niche service providers.

The principal reason why seismic data are important, especially during oil exploration, is because of the wide geographical area they cover. A survey may cover several square kilometers. Seismic thus provides a much-appreciated overview of an area at different depths. The Norwegian continental shelf, as shown in figure 3.3, is thoroughly surveyed by seismic.

The fundamental challenge during exploration is the *search*. Explorationists look for a pattern of conditions that enable the generation of hydrocarbons:[3] a source rock (containing the past vegetation), a migration path (channeling the hydrocarbons), and a trap (accumulating hydrocarbons in an oil reservoir; see chapter 2). The hydrocarbons in the North Sea are contained in Jurassic sandstone buried by layers of sediment deposited over millions of years. In seismic interpretation the challenge is to tie the contours generated by seismic, based on visualized color conventions to differentiate

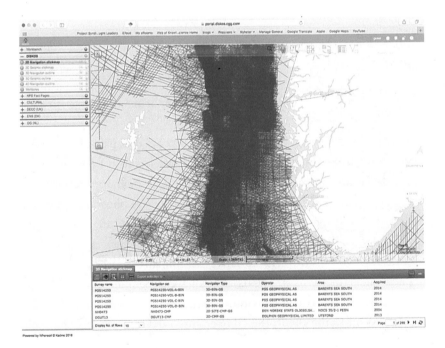

Figure 3.3
Map of seismic surveys on the southern part of the Norwegian continental shelf. The surveys (*blue line*) are so dense in large areas that they fill the map.
Source: Reproduced by permission from the Norwegian Petroleum Directorate.

differences in the rate and speed of seismic reflections, to geological rock formations, known as a *horizon*. As analyzed in more detail in chapter 5, this search takes a number of forms, depending, among other things, on the availability (or not) of data types other than seismic.

There are, however, several issues with seismic data. Most importantly, seismic data are relatively coarse, with granularity in the neighborhood of a cube with sides of one hundred meters—"about the size of this office building" as explained by one geologist. Given the importance of local geological processes, such as faults, the poor resolution of seismic data inevitably implies that potentially relevant and consequential geological circumstances go unnoticed.

In addition, certain areas are in a seismic "shadow." For instance, at Ekofisk, the oil field that initiated Norwegian oil in the 1960s, a gas-obstructed crestal area covering about one-third of the oil field effectively blocks seismic signals.

Recently, selected offshore oil fields have been instrumented to capture 4D seismic—that is, 3D seismic in a time series—typically every six months. 4D seismic is also known as permanent reservoir monitoring (see figure 3.4). The seabed is instrumented with thousands of networked sensors. It comes with a significant investment. The business value is plausible but has not been compellingly demonstrated to date, hence investments have been moderate. On the Norwegian continental shelf, there are currently three fields with 4D seismic (Ekofisk, Grane, Osberg). The promise of 4D seismic is primarily tied to the possibility of improving oil recovery during the later stages of an oil field's life cycle. After an initial, relatively brief period during which the pressure in the oil reservoir is sufficiently high to push hydrocarbons out (given a drilled well), later stages rely on recreating the reservoir's pressure by injecting water (or gas) into the reservoir from supplementary *injection* wells. 4D seismic, then, represents a method to trace the waterfront as it moves over time inside the oil reservoir from injection to production wells. Late-stage oil fields have an increasingly high *water cut* (i.e., the fraction of water mixed in with the flow of produced hydrocarbons). For instance, Ekofisk, the oldest field, "produces more than 50 percent water," as one production engineer explained. 4D seismic is central to ambitions of increased oil recovery. With

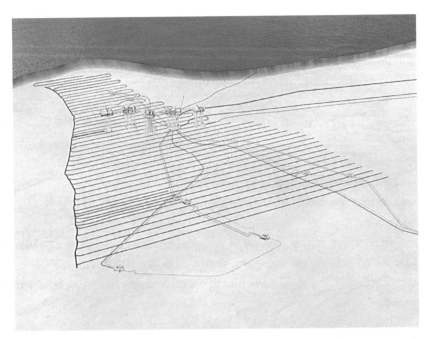

Figure 3.4
Permanent reservoir monitoring of the Johan Sverdrup field. A network of seismic sensors cover about 120 square kilometers of the seabed.
Source: Reproduced by permission from Equinor.

global recovery rates historically at little more than 25 percent, the present average on the Norwegian continental shelf is 46 percent. The goal at Johan Sverdrup, the latest oil "elephant" (an oil discovery exceeding five hundred million barrels) is 70 percent with 4D seismic through sixty-five hundred sensors spread out over 120 square kilometers with 360 kilometers of fiber-optic network (Hovland 2018). The massive amounts of data generated from 3D seismic are multiplied with 4D. The work of interpreting the seismic, however, struggles to keep up with the pace of new data. As one geologist complained, "[4D seismic produces] more than we can digest." Thus, 4D seismic, driven by the broad push for data science–based, data-driven practices, runs up against the all-too-familiar challenges of making *actual* use of the data in work practices of exploration (see chapter 5).

Drilling   The data from drilling operations are IoT measurements from the sensors installed on the drill bit and along the drill string while drilling a new well. The sensors measure pressure, temperature, and torque (the force of the rotating drill pipe). In addition, physical samples of rock cuttings are transported up to the platform deck by the circulation of the lubricating drill mud (a mix of bentonite clay and barium sulphate metals to increase the weight), as well as, when finally hitting the oil reservoir, a core sample. The physical samples are archived by the Norwegian Petroleum Directorate (NPD) and are not privately owned by the oil companies. The archive of core samples dates back to 1965 (see figure 3.5).

Drilling is a hazardous operation. Risks of smaller incidents, such as having a stuck pipe that subsequently gets snapped off or collapsing walls in the well, are immanent, as well as more serious ones such as a *kick*, when the pressure in the reservoir, usually balanced by the weight of the column of drilling mud injected into the well for lubrication when drilling into rock formations, is underestimated. This results in the flow of hydrocarbons into the well. A blowout, the ultimate risk during drilling, as the 2010 Deepwater Horizon accident vividly demonstrated, is an uncontrolled kick.

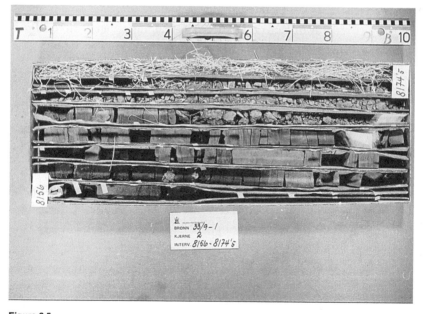

Figure 3.5

A core sample.

Source: Reproduced by permission from the Norwegian Petroleum Directorate.

The communication bandwidth from the downhole sensors of the drill bit to the drillers on the platform and the onshore control room operators is poor. Communication is via physical pulses propagated by the drill mud, akin to Morse code. Even with more recent improvements, "mud-pulse telemetry rates have improved to more than 20 bits/sec (bps) at depths shallower than 20,000 ft, and in excess of 3 bps from depths of more than 36,000 ft. In 1978, a typical data rate was 0.4 bps" (*Oil & Gas Journal* 2008).

Drilling data are likewise sparse. Drillers operate on crude outlines or trends from the sensors rather than from accurate, updated readings. Shifts toward more data-driven practices are also a challenge to drilling, the most conservative (due to safety regulations) of oil activities. One initiative is wired drill pipe (WDP). It involves installing a broadband communication channel along the drill string, allowing a ten thousandfold increase in capacity.[4] The implementation of WDP, however, has been slow. A study of a pilot on the Norwegian continental shelf found that drillers, instead of exploiting the increased granularity and timeliness of the sensor readings, filtered out the additional information, as if it were noise, to recreate familiar, crude trends (Hjelle 2015). This reiterates the old adage that data do not in any straightforward way drive action or decision-making in digital oil (Feldman and March 1981).

There are several issues with drilling data. First, to the frustration of explorationists, offshore, deep-reservoir drilling is costly even by the inflated standards of oil. A drilling campaign takes a month and could run to USD $100 million. While exploring for oil (see chapter 5), you accordingly make the most of the available seismic and other data. Yet drilling provides the proof is in the pudding for whether or not your geological understanding is accurate. "I waited several years," one explorationist said, "before my prospect [candidate for an oil find] was drilled." The feedback loop from an exploration project, yielding an educated guess regarding an oil discovery (Fine 2009; Pollock and Williams 2010), spans years, if drilled at all.

An example is the most recent big oil discovery (an elephant), Johan Sverdrup. In the account by the NPD, with due emphasis on the importance of reinterpreting the public good of historic data maintained in the Diskos geobank by NPD (see chapter 2), the discovery was attributed to reinterpreting "two old Statoil wells in the area (16/1–14 and 16/1–15). The first contained gas-fractured basement rocks, while the other contained oil traces on top of a 250-metre good sandstone column." The data were then combined with new seismic and new geological models for "unproven plays" (see more in chapter 5).

Second, the data quality of sensor measurements from drilling is, at best, varied. Given the cost, drilling through the overburden (the upper layers of a geological formation), especially during the first couple of decades of Norwegian oil in the 1970s and 1980s, is a chore to be done as quickly as possible. Precious little energy is invested in ensuring high data quality from the overburden. More recently, the shallower sedimentary layers have attracted geological interest. For instance, there is increased interest in subtle traps with smaller but still commercially interesting quantities of hydrocarbons and in the Barents Sea, with reservoirs at more shallow depths than in the North or Norwegian Sea. These efforts, however, are hampered by the questionable quality of old drilling data, with the dominant sentiment being "We just wanted to quickly get to the deeper levels."

The physical wear and tear on downhole sensors is tremendous. They regularly stop working or are off calibration. As a response, several oil operators have established twenty-four seven onshore control rooms to monitor data quality during drilling campaigns. Sitting in a room surrounded by screens displaying real-time drilling data, one control room operator explained how "I call up the drillers immediately when I receive strange readings and ask them to check." In addition, he explained, "We run scripts [i.e., simple algorithms] that automatically check that the sensor values we receive are within pre-defined thresholds." Hence, manual and (semi-)automatic data quality procedures are in place for drilling data.

Well logging   Data from the logging of wells are obtained by lowering a collection of sensors into the well. Pioneered by the founders of the oil-service company Schlumberger, which lowers electrodes into the well, the formative idea is simple: "If you get a strong [magnetic] field, you have been moving through a highly conductive formation—typically water. If you get a very weak field, they you have been moving through a highly resistive formation—oil or granite or something else" (Bowker 1994, 5). The sensors measure physical properties indicative of geological characteristics or rock types surrounding the well at the depth to which the equipment has been lowered. An assembly of sensors would normally measure electric (resistivity), electromagnetic, acoustic, and radioactive (gamma radiation) properties. A specific measurement or a combination thereof is associated with certain geological characteristics. For instance, gamma radiation is higher in shale than in sandstone, and electrical resistivity is higher in oil than in water. The log plots different measurements and observations along the well along a downward axis representing the depth of the well. The result of well logging is a wire line log, indicative of a lithographic column of the layers of rock (see figure 3.6).

Well logs are usually generated as part of scheduled halts of operations in connection with maintenance, calibration, or equipment testing. More recently, well logging while drilling (LWD) has been attempted. Potentially, the additional real-time data from LWD could help drillers navigate more

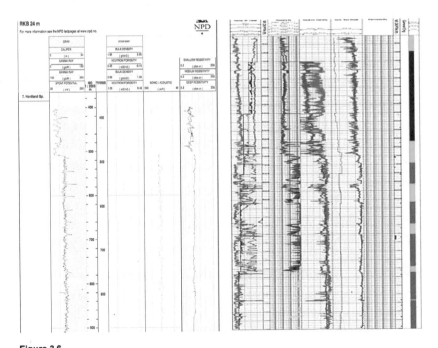

Figure 3.6
A well log.
Source: Reproduced by permission from the Norwegian Petroleum Directorate.

accurately into the geological formation, making it safer and more efficient to drill and to ensure equipment is appropriately placed within the oil reservoir to maximize production.

Production   Producing oil fields generate a steady stream of real-time sensor data. Time-stamped every few seconds, pressure, fluid flow rate, temperature, vibration, composition, fluid holdup, and electromagnetic resistance readings come from sensors located along pipelines and at the gauges, chokes, and separators of the production facility (see figure 3.7). One subsea well will typically have about five to ten sensors. Traditionally run from offshore platforms, more and more tasks (and manpower, often against the will of the unions) have been shifted onshore. Production engineers, intently observing their screens with IoT-based production data, are tasked with the short-term optimization of production by juggling the chokes, gauges, and lifts controlling a producing well. The short-term focus of production engineers is interleaved, in some oil companies by creating shared workspaces, with the long-term focus of reservoir engineers. If production engineers look for local production optimization in one well, a reservoir engineer is looking for production optimization across a considerable collection of wells drilled in the field and planning the placing and timing of later production and/or injection wells to ensure adequate overall performance of the field as a whole for its full life cycle. A reservoir model, a finite-element, 3D-simulation model, is crucial. The reservoir model, "requiring several months of work," is, as one geoscientist noted with a smile, "never right," consistent with observations of large, complex models in science (Millo and MacKenzie 2009; Edwards 2010; Lahsen 2005; Sundberg 2010). Striving but never

DATEPRD	WELL_BORE_CODE	DOWNHOLE_PRESS	DOWNHOLE_TEMPER	AVG_DP_TUBING	FLU	AVG_CHOKE_SIZE_P	AVG_WHP_P	AVG_WHT_P	DP_CHOKE_SIZE	BORE_OIL_VOL	BORE_GAS_VOL
25.apr.14	NO 15/9-F-1 C	255,527	107,971	165,539		48,53377 %	89,988	64,547	61,405	1 249	178 064
26.apr.14	NO 15/9-F-1 C	247,199	108,052	162,422		49,84450 %	84,777	65,724	56,148	1 346	192 602
27.apr.14	NO 15/9-F-1 C	240,736	108,054	158,899		50,29698 %	80,857	66,934	52,202	1 350	194 496
28.apr.14	NO 15/9-F-1 C	235,021	108,042	157,683		50,73588 %	77,338	67,848	48,708	1 846	192 900
29.apr.14	NO 15/9-F-1 C	232,744	107,988	156,795		50,11392 %	75,949	65,707	47,376	1 279	184 900
30.apr.14	NO 15/9-F-1 C	233,298	107,893	157,179		48,92679 %	76,118	62,796	47,611	1 226	177 108

Figure 3.7

Production data sampled from a few days in April 2014 from the Volve field. The parameter registers include downhole pressure, downhole temperature, average choke size, average weight of mud, bore oil volume, and bore gas volume.

Source: Reproduced by permission from the Volve open data set.

achieving perfection, the reservoir models are continuously calibrated based on the real-time production data streams, a task known as *history matching*.

Monitoring and maintenance Maintenance through the monitoring of operations is gradually shifting focus from traditionally reactive, after-the-fact interventions to prescriptive, preemptive condition-based maintenance. This implies that maintenance is shifting toward more data-driven predictive interventions based on real-time IoT and historic data. This shift mirrors a much broader shift across numerous domains, including predictive policing (Waardenburg et al. 2018), marketing (Alaimo and Kallinikos 2018), and forensic evidence (Harcourt 2008).

Well maintenance (*interventions*) mainly repairs or replaces selected components of the subsea well. Wells can fill with sand and require "washing" by injecting chemicals into the well at a designated pressure. Many subsea well interventions are due to *scale*. Scale consists mainly of inorganic salts that have elements of calcium carbonates, barium, and strontium sulphates. The production tubing gets clogged from scale that severely hampers the flow of hydrocarbons in the well. Scale typically develops when reservoir formation water (i.e., water contained inside the reservoir) enters the well. When the formation water undergoes changes in pressure and temperature, or when two incompatible fluids intermingle, either sulphate or carbonate scales may develop. Even relatively new subsea wells may suffer from scale if the drilling or completion fluid is incompatible with the formation water. During production, as oil and gas are gradually drained, increased amounts of formation water are produced together with the hydrocarbons and are likely to lead to challenges with scale.

Well intervention also involves replacing or upgrading faulty or outdated components. Well temperature and pressure transmitters tend to have short life cycles. They are, however, expensive to replace when this entails shutting down the well.

Maintenance during production is costly. Not only are significant tasks contracted to service companies; maintenance also regularly involves temporarily reducing or shutting down production. Oil operators, NorthOil included, in an effort to reduce the significant cost of maintenance, have explored ways to standardize an interesting area of maintenance operations known as *lightweight interventions.*

The key challenge facing light-weight interventions is that the envisioned efficiency gains from economy of scale (standardization) run counter to the uniqueness of every well. As pointed out in the Obama administration's National Commission on the BP *Deepwater Horizon* Oil Spill and Offshore Drilling (2011), "Every well is unique." First, as geologists would quickly underscore, the geology is, literally, local. A central challenge when developing a geological understanding of a particular area is to interleave an emphasis on continuity ("This basin area consists of a given sequence of sedimentary layers") with a deep appreciation of the many variants of local, idiosyncratic geological processes that make every location unique. Compounding the local nature of the geology, the technical configuration of production facilities is bespoke. As we note in an earlier study (Monteiro et al. 2012b, 175), "You need to know the personality of the well [to do maintenance/intervention]." Beyond the uniqueness of geological sites, the technical configuration and detailed processes of drilling and completion (casing or cementing the walls of the well) are unique to each well. Thousands of documents are produced for a single well, detailing the position of the well, the subsea equipment installed, the diameter of the well at different depths, the well completion method, and, not the least, the experience reports from the drilling and operation of the well.

Routine monitoring is not only geared toward maintenance but is also very much about controlling the inherent risks to human life, economic value, and the environment. A notable aspect of such monitoring is the detection of sand as part of the produced flow of hydrocarbons. Sand erodes chokes, valves, and pipes and hence may wreak havoc. Given that many of the hydrocarbons in the Norwegian continental shelf are found in Jurassic sandstone, sand is an all-too-real risk, as discussed in detail in chapter 6.

The oil industry is under increasing pressure. Concerns about climate change resulting from our reliance on fossil fuels are significant and mounting. In addition, concerns regarding hazards to the environment from pollution, oil spills, or regular operations have infiltrated Norwegian oil activities from the very beginning. With the oil industry proactively lobbying to open presently banned areas in the Arctic part of the North Sea and the high north in the Barents Sea, oil operators are starting to monitor the marine environment in an effort to present themselves as (more) environmentally conscious. Distinctly different from monitoring sand, a well-defined phenomenon, monitoring the marine environment immediately prompts questions of what aspects of the radically open-ended phenomena of a marine environment are to be monitored, with candidates including but not limited to fish, algae, flora, water quality, and benthic, seabed sediments. What aspects of the marine environment are relevant to capture and, equally important yet often implicit, for whom and why? And what aspects are technically, practically, and economically feasible to capture with IoT-based marine environment monitoring? This was the situation facing NorthOil's efforts in marine environmental monitoring, detailed in chapter 7.

As should be evident from the above, work practices across all phases of oil—exploration, drilling, production, logging, maintenance, and monitoring—produce and rely heavily on data. In the slogan-like vocabulary of big data, four Vs characterize the data: volume, variety, velocity, and veracity.[5] Data in digital oil clearly exemplify these four Vs. The volume of data is considerable, with seismic data alone accounting for one terabyte for every survey and production data accumulating over decades. The variety of data is significant. In addition, and as a healthy antidote to inclinations toward "size is everything," some data sets are small in volume but carry a disproportionally heavy weight—notably, slide presentations providing much-needed overviews and summaries. Velocity comes from the ever-increasing availability of real-time data from production and monitoring. Sensors are notoriously unreliably, noisy, or off calibration, so the veracity of the data is chronically a concern. As one informant explained, "We've given up fixing that choke. It's too costly to replace. I tell [the production engineer] that he simply has to shut his eyes and disregard the readings from it" (Monteiro et al. 2012a, 101).

Wary of the forceful rhetorical and ideological qualities of the ongoing advocacy for big data, subsequent chapters analyze the extensive set of practices implicated in *doing* data-driven digital oil. The combination of high variety and low veracity is, for instance, characteristic of big digital oil data, more so than size alone. It prompts a varied and elaborate set of strategies to triangulate, quality check, and, ultimately, accumulate credibility for the data—enough credibility for action and decision-making.

SILOS: BARRIERS TO COMMUNICATION

From what I have outlined, digital oil is awash in data; digital oil exhibits all the characteristics of big data. The everyday practices of knowing digital oil, however, grapple with data friction that makes accessing, interpretating, and sharing geodata anything but frictionless. The institutional, political, and technical barriers to the effortless flow of digital oil data stem from the circumstances in Norwegian-based offshore oil outlined in chapter 2. These barriers give rise to the different types of geodata silos I go on to elaborate.

Business boundaries   The oil industry, as well as Norwegian offshore oil, which is the focus of this book, has a level of outsourcing significantly above many, if not most, other industries. Quite different from the monolithic organization of production pioneered by Henry Ford, which controlled all phases of production including those of the key input resources, such as growing the company's own rubber on Brazilian plantations (Staudenmaier 1997), the oil industry is organized into an industrial *ecosystem* with a number of independent organizations (see chapter 2). Also, seemingly central tasks are performed outside—that is, outsourced or contracted by the oil operator. The oil operator largely focuses on the upstream activities of exploration and production. Drilling is conducted by separate businesses. Global market leaders include Halliburton, Transocean, and Schlumberger. The drilling rigs are contracted. Large oil-service companies such as Schlumberger also offer services in production optimization and seismic interpretations, services intimately tied to key concerns of the oil operator. During fieldwork for this book, I and my fellow researchers would regularly have a hard time separating informants belonging to the oil operator from those with the service companies. They would often share offices and apparently work seamlessly together, regulated by long-term contracts. Seismic surveys, too, are generated by specialized organizations using specialized vessels. Rig and production facility construction is contracted. Maintenance and intervention are largely conducted by contracted companies. Providing supply vessels and platform anchor handling are contract work. The many different organizations create challenges for collaboration and communication; the industrial ecosystem of upstream oil struggles with organizational silos.

The splitting of activities across distinct business organizations, each with its own priorities, creates challenges for collaboration, communication, and information sharing (Carlile 2004; Star 2010). The outsourcing of drilling is illustrative. Given the cost of hiring a drill rig and crew, the contracts regulating the drilling campaigns focus on the speed of drilling—that is, minimizing the time to reach the (assumed) oil reservoir. As one geologist complained, "The drillers don't care about where [in the reservoir, important for long-term production efficiency] they hit as long as they're fast." Drilling contracts are normally by the day, hence incentivizing speed over quality.[6] Each drilling company has its own equipment, which means that "the measurement [data] you receive from one of the drilling companies cannot easily be compared with another." As Bowker (1994) noted in his historical reconstruction of Schlumberger from the 1920s through the 1940s, oil-service companies have a vested interest in keeping the details of their measurements and methods black boxed:

> What mattered here was creating a space within which Schlumberger could work—a space into which flowed equipment and resources and out of which flowed curves and their interpretations, but within which things were kept murky so that no one could understand the full process and so that Schlumberger could retain control of the process of interpretation. (153)

An example of this black boxing is the multiphase meter allowing oil and gas to share pipeline infrastructures and pumps. The measurements are aggregated and manipulated data, not the underpinning measurements, which remain only with the vendor and service company, not the oil operator. The oil-service companies' black boxing of the raw data of their measurements thus undermines the possibilities of data science: data-driven repurposing of the data. The configuration of the industrial ecosystem, all jockeying for the dominant role of platform owner (see chapter 1), has clear elements of a power struggle. As one oil-operator geoscientist, expressing frustration, put in his own colorful words, "[Large oil-service companies] have us by the balls."

Historized boundaries　Digital oil data are *historized* in more than one sense of the word. In an age in which we all resort to search engines, "googling" by now is an everyday phrase in many languages, Norwegian being no exception, so why are geoscientists unable to search and find relevant information with the same ease as on the web? After all, the size of geodata is dwarfed by the size of the web, which is what search engines index. Why, then, do geoscientists struggle to access, find, and make sense of digital oil data? The answer, in short, is that the web, despite its vastness, is "flat" in ways digital oil data are not; digital oil data have a topology. Let me elaborate on this perhaps counterintuitive claim.

Given the relative lack of attention to old drilling data from the initial, shallow parts of a drilling campaign, the dating of data is crucial to assess its relevance. Moreover, naming conventions for wells and installations vary over time and across fields and professional subdisciplines. As we found in an earlier study (Hepsø et al. 2009), "If you didn't follow the well from its inception, there is no way you can know where to find the information or what kind of information that is available. Thus, it is also impossible to just use the search engine."

Moreover, digital oil data are subject to an elaborate regime of role- and situation-based access rights depending on one's position and (transient) project. An information search is thus relative to access rights.[7] As we found, "Sometimes you do not find information just because you do not have access . . . so you have to call various people and ask . . . it is very time consuming and I know some people do not bother spending all their time on that . . . however, not having important information means more uncertainty during operation, and this can increase the risk and cost of operation" (Jarulaitis and Monteiro 2009, 9).

Making data storage, access, and retrieval more frictionless has been high on NorthOil's agenda for a long time. During the last couple of decades, there have been two strategically promoted initiatives to migrate from the initial data platform to a later one. Despite heavy investments, each of these two efforts to fully migrate the data from one generation of the platform to the next resulted in just a partial migration. Discovering halfway into the migration that it had underestimated the many ways data were entangled with the platform, NorthOil layered the partially migrated data on top of, instead of substituting for, the data tied to the previous platform. Information access and retrieval then is now across generations of platforms holding data, not entirely inside any one platform. As we found in our study (Jarulaitis and Monteiro 2010), NorthOil was "not able to fully migrate" from the first-generation data platform to the next, hence:

> This [data on the first-generation platform] lived on together with [the second-generation platform]. . . . Later we got [the third-generation platform] . . . and then [the first-] and [the second-generation platforms] still lived on because it was *impossible to migrate* with all the historical data we needed. When you need it [the historical system(s)] you can always add some new information to it. . . . [smiling]. So now you have [the first-], [the second-] and [the third-generation platform] . . . when something new comes [after the third-generation], we will probably still keep those three old ones [smiling]. (Manager responsible for operational support, 7; emphasis added)

This situation with, effectively, historically stratified data platforms led to numerous workarounds to help frustrated users locate specific data and documents:

> Sometimes I get a call in the evening from offshore people saying that they have been searching for a specific document for an hour or so with no success. . . . To avoid this we have developed a practice that for every new drilling program, a drilling engineer [working onshore] creates an excel document containing links to documents that are the most important ones for drilling engineers working offshore. It is additional work as we [engineers working onshore] have to update those excel documents during drilling, but then offshore people have much better overview. (Drilling engineer working onshore)

Professional boundaries The subsurface community of geoscientists is anything but homogeneous. It consists of numerous subdisciplines, the most important being geologists, geophysicists, petrochemical engineers, reservoir engineers, production engineers, and lithographic loggers. Obviously, an overgeneralization but still vividly felt,[8] the subdisciplines are tasked with specialized problems that limit communication, collaboration, and sharing across the subdisciplines. Importantly, the organizational and professional silos get reinforced by the lack of integrated digital tools. There is a proliferation of specialized, niche digital tools, each supporting specific subdisciplines (see table 3.1).

In sum, professional boundaries within the subsurface community of different geoscience disciplines are upheld by differences in their *vocabulary*, the *data types* they focus on (see the previous section outlining the phases of exploration, drilling, well logging, production, and maintenance), their *digital tools*, and their *gaze*, understood as their *temporal focus*. For instance, production engineers operate with a real-time (hours, days) focus, process engineers operate with a horizon of a few years, and geologists operate in time frames of millions of years (Goodwin 1994; Mol 2003).

RESPONDING TO SILOS: DIGITAL DATA PLATFORMS AND INDUSTRY-WIDE STANDARDIZATION EFFORTS

The dangers of trapping geodata in silos, thus undermining opportunities for sharing and collaboration, were explicitly targeted during the formative years of Norwegian oil regulation in the late 1960s and early 1970s. The promotion of "open" Norwegian oil data accordingly started early, long before the concept of open data attracted the kind of interest it has sparked recently. As outlined in chapter 2, Norwegian authorities imposed regulations to facilitate data sharing, effectively striving toward making geodata a public good (Ostrom 1990). Distinctly different from the norms in the rest of the world, including the US, the North Sea data were made public:

> The measurements [i.e., the IoT-based data] are conducted privately and held separately, field by field, by rival companies, or by oil service firms contracted to private or national oil-production companies. . . . In fact, *apart from the British and Norwegian zones of the North Sea, there is no production region in the world for which field-by-field production data is publicly available.* (Mitchell 2011, 245; emphasis added)

The open data at the NPD are comprehensive. The NPD (2020) contains "all" the geodata generated throughout Norway's fifty years of oil. It contains seismic, well logs, and production data from practically all current and abandoned oil fields.[9]

There is, however, a significant discrepancy between the ambition of making all the data available and the possibilities, in practice, of doing exactly that. The NPD data are anything but frictionless. Available, in principle, at your fingertips, in practice data are of little value without additional, contextual data about the circumstances under which the data were captured and, not the least, the degree of associated uncertainties in the data. The data in the NPD, in other words, have for all purposes not been sufficiently sanitized and "washed."

The limitations and challenges of the NPD data, which undermine the data's status as open, are well known. An influential industrial consortium comprising most companies in the Norwegian industrial oil and gas ecosystem points out how crucial it is "to achieve increased quality, productivity and efficiency across the whole value chain [in oil and gas] by regulated sharing and use of datasets between the companies" and, furthermore, to "collaborate on the introduction of regulation and standards for shared solutions for storing, sharing and use of data throughout the whole value-chain," with digital platforms identified as the principal enabler (Konkraft 2018, 10).

Alongside these initiatives have been several attempts to platformize vendors' tools (see Plantin et al. 2018)—that is, transform earlier stand-alone systems into digital platforms with associated, evolving ecosystems, thus tapping into the logic of network externalities to break off from silos (Gawer 2011; Parker and Van Alstyne 2014; Tiwana 2013). For instance, OSIsoft's PI System enjoys this position, as it is used as a "historian" into which all production data are streamed. In seismic interpretation, geological modeling, and reservoir modeling, Schlumberger (2021) has a dominant position. Through organic and nonorganic acquisition, it offers a comprehensive ecosystem of tools. The lock-in effect of the network externalities is strong. As one North-Oil geoscientist explained, "We try to break out of their grip." Alternative, including open source–based, software platforms exist that cover many of the areas, but their market share remains modest.[10]

Finally, there *are industry-wide efforts at standardization*, with the subsequent establishment of a digital platforms ecosystem for data. An important example is Open Subsurface Universe Data (OSDU) and its attempt to platformize subsurface geodata (wells, logs, reservoir, maintenance).[11] As pointed out in a recent industry consortium white paper, the practical usefulness of the open geodata leaves much to be desired, triggering renewed efforts into opening up proprietary data through industry-wide standardization. Similar industry-wide standardization and platform formation for other parts of the oil and gas value chain exist, such as for manufacturing and the engineering of equipment.[12]

CONCLUSION

The broad trends toward advocating more data-driven practices strongly influence the ongoing digital transformation of the industrial ecosystem of offshore oil and gas on the Norwegian continental shelf (NCS). However, it is crucial to unpack the characteristics of the data underpinning the data-driven work practices of digital oil. This chapter has highlighted the salient features of digital oil data. They come in several different types of data. In digital oil, size is not everything. "Small" data—for instance a slide presentation supporting a geological interpretation—carry disproportional weight. IoT-based data dominate. Notoriously error prone and off calibration, there is no alternative to IoT-based data, hence the issue is one of devising strategies to cope. Data are historized and thus profoundly shaped and formatted by the conditions and circumstances surrounding them. Data are captured for particular purposes that create friction, with later attempts at repurposing the data. Data are trapped in silos as they become tightly coupled to niche-oriented digital tools supporting each of the many distinct professional and disciplinary communities within an oil operator such as NorthOil. Not only within NorthOil but also across the whole industrial ecosystem of subcontractors, service providers, and technology vendors are institutional barriers limiting access to data. Several actors and industry-wide initiatives are attempting to establish and control the digital platform supporting the ecosystem of digital geodata. The NPD's open geodata, institutionally emerging from Norway's history of considering the sources of energy production (waterfalls, the continental shelf, wind) as public goods, should be understood as an attempt to establish a publicly controlled platform for geo-data. It is, however, struggling with issues of quality, and hence trust of the data, compounded with cumbersome access and navigation, which severely undercuts its role as an open infrastructure (Frischmann 2012). With the present chapter completing the necessary backdrop, the subsequent empirical chapters in part II of this book set out to detail, analyze, and discuss the nature and implications of practices of knowing digital oil.

II CASES

4 DATA

written with Marius Mikalsen

No data are truly "raw."
—Bateson (1972)

One should never speak of "data"—what is given—but rather of *sublata*,
that is, "achievements."
— Latour (1999)

The rising role and presence of data-driven decision-making based on predictions are empirically visible across wide and varied domains and settings (Shrestha et al. 2019), including but not limited to medicine (e.g., diagnosing skin cancer; Esteva et al. 2017), computer vision (Krizhevsky et al. 2012), security (e.g., predictive policing; Waardenburg et al. 2018), finance (e.g., credit risk assessment; Pacelli and Azzollini 2011), transport (e.g., autonomous vehicles; Hoogendoorn et al. 2014; Chen et al. 2015), and human resource management (van den Broek et al. 2021). Within process industry and manufacturing, which share many aspects with the domain of offshore oil and gas covered by this book, there is a push toward visions of Industry 4.0 and Industrial Internet of Things (IIoT) known as *condition-based maintenance* (CBM): predictions fed by IoT measurements of temperature, vibration, noise, pressure, and other details of the physical conditions of the production equipment and components. Maintenance of equipment and installation, previously part of routine monitoring, supplemented by reactive measures in response to incidents, is thus gradually being substituted with a preemptive strategy fed by IoT data from the physical components. Direct inspection of the physical conditions of the equipment is hence turned into an algorithmic phenomenon (see chapter 1) to be interpreted by maintenance operators. As de Jonge et al. (2017) point out, however, there is presently a lack of research analyzing the social and practical circumstances for the uptake of the predictions for wear and tear generated by condition-based monitoring algorithms.

A domain particularly strongly influenced during the last decade by the rise of data-driven data science is marketing and advertising. First in the US but later in the UK and somewhat later in the rest of Europe, marketing has been transformed by what is known as *programmatic advertising* (Alaimo and Kallinikos 2018). Programmatic advertisement is predicting consumer preferences based on profiling trace data from internet search histories, website cookies, and Facebook postings and likes, as well as other sources (Gerlitz and Helmond 2013).

The effectiveness of employing online behavioral traces to "nudge" consumers has demonstrated the potential scope and reach,[1] especially when supplemented with additional data types, of the data-driven manipulation of a wider spectrum of user/consumer/citizen behaviors. There is growing concern over the amount of data traces we leave, not only on the net but, increasingly, in all walks of life, such as our physical location, tracked via GPS by our ever-present mobile phones; our physical conditions, such as heart rate, respiration, and sweat captured by Fitbits; our economic transactions, given the dwindling presence of cash; and our sexual, political, and psychological profiles from engagement with apps (Duportail 2017). A steady stream of scandals has fueled public outcry. The list is long and growing, with notable examples that include Edward Snowden's abundantly clear demonstration of the scope of the US National Security Agency's surveillance program to filter (i.e., predict) subjects, Cambridge Analytica's access to Facebook users' profiles, and the Chinese surveillance regime of the Muslim Uyghur minority population.

Few, if any, have delivered a more scathing critique, with a comprehensive analytic diagnosis of the underlying dynamics, than Zuboff (2019). Her notion of surveillance capitalism captures how, far from singular glitches, as Facebook's Mark Zuckerberg claimed in a US Senate hearing (*New York*

Times 2018), there is a systematic exploitation of what Zuboff calls the *behavior surplus* of data traces we produce. We are seduced by attractive services, and most often without our informed consent, a growing set of data traces from our everyday behavior is collected, aggregated, combined, sold, and resold by powerful if not monopolistic tech companies. This corresponds closely to Bauman's (2007) observation that a defining aspect of consumerism is how consumers are transformed into commodities.

In an early and provocatively formulated vision of the consequences of the expanding reach of data-driven approaches, Anderson (2008; emphasis added) outlined an extreme "no theory" position: "*Out with every theory of human behavior*, from linguistics to sociology. Forget taxonomy, ontology, and psychology. Who knows why people do what they do? The point is they do it, and we can track and measure it with unprecedented fidelity. *With enough data, the numbers speak for themselves.*" Using the radical influence of data-driven approaches on marketing as leverage, Anderson (2008) explicitly went on to proclaim that "the big target here isn't advertising, though. It's science," with the result that the "[traditional] approach to science—[i.e., to] hypothesize, model, test—is becoming obsolete." Vast, increasingly rich data tethered to methods and algorithms from machine learning generate the predictions that models/theory once occupied. Correlation, not theoretically modeled causality, the argument goes, is king.

A no theory vision is not confined only to outlets like *Wired*, where Anderson's (2008) proclamation appeared. In *Science* (Lazer et al. 2009) and *Nature* (LeCun et al. 2015; Watts 2007), too, albeit with considerably more attention to hurdles and preconditions (cf. Leonelli 2019; Rahwan et al. 2019), the potential to shape scientific practices and tools by employing data-driven approaches are discussed.

However, the no theory vision is, rightly, recognized by critically oriented scholars as utopian. Kitchin (2014, 3) identifies the promise of a fourth paradigm of science as "Big Data ushers in a new era of empiricism, wherein the volume of data, accompanied by techniques that can reveal their inherent truth, enables data to speak for themselves free of theory." He then goes on to explain that the vision is flawed because "all data provide oligoptic views of the world: views from certain vantage points, using particular tools, rather than an all-seeing, infallible God's eye view"; hence, "whilst data can be interpreted free of context and domain-specific expertise, such an epistemological interpretation is likely to be anaemic or unhelpful as it lacks embedding in wider debates and knowledge" (5). Similarly, van den Broek et al. (2021) argue that the vision of a purely inductive, data-driven science "represents a radical empiricist mode" of knowing (7). The emergent, motley field of critical data (science)/algorithm studies—spanning across a number of disciplines but notably science and technology studies (STS)/knowledge infrastructures (Bechmann and Bowker 2019; Edwards et al. 2011; Henke and Sims 2020; Jackson 2014; Ribes and Polk 2015; Ribes 2019), sociology theory (Glaser et al. 2021; Hà and Chow-White 2021; Plantin 2019; Seyfert and Roberge 2016), information systems (Parmiggiani et al. 2021; van den Broek et al. 2021), and law (Lehr and Ohm 2017)—could be understood as defined by a rejection of a no theory vision.

Interestingly, the critique of a no theory position is not confined to where you would expect—namely, critical, socially informed studies of data science. The critique also comes from deep within the data sciences themselves. This underscores the importance of treating data-driven data science not as a monolith (see Sugimoto et al. 2016) but as a motley of distinct approaches yielding a nuanced and varied phenomenon of data science (Shrestha et al. 2019). One notable voice is Judea Pearl, 2011 winner of the Turing Award, which in computer science comes closest to the Fields Medal in mathematics or the Nobel Prize, for his pioneering work with one method in the data sciences, Bayesian networks. The epitomized expression of a no theory approach in data science is neural networks in general and deep learning in particular. Deep learning has for the last few years produced a series of stunning results in computer vision: driving a car using only cameras (Bojarski et al. 2016), diagnosing skin cancer (Esteva et al. 2017), and, when combined with reinforcement learning, beating human champions at the game of Go and surpassing all previous approaches in chess (Mnih et al. 2015; Silver et al. 2016). Still, Pearl and MacKenzie (2018) argue, deep learning will never do more than mathematically sophisticated "curve fitting":

They are driven by a stream of observations to which they attempt to *fit a function, in much the same way that a statistician tries to fit a line to a collection of points.* Deep neural networks have added many more layers to the complexity of the fitted function, but raw data still drives the fitting process. (30–31; emphasis added)

A healthy skepticism to proclaimed implications of technology in general and data science in particular clearly is necessary. Proclamations of no theory are unwarranted. But what is missing in the scathing critique of the visions of data science by Kitchin and others is an account about how, despite sound theoretical objections, there are numerous empirical situations in which the repurposing of data and simplistic analytics "work." Big data, to paraphrase LaPorte and Consolini (1991), might not work in theory, but it does work—in certain configurations, for selected purposes, for some situations—in practice. There is, accordingly, a need to look closer at the empirical conditions and practices implicated in *making* it work.

THEORETICAL FRAMING: DATA WORK AND REPAIR

As argued in chapter 1, a crucial analytic and empirical shortcoming of a data-driven no-theory vision is the radical underappreciation of what goes into the making of data. Traditionally, a representational view has data corresponding directly with some given, preexisting physical object, process, or quality. Such a view, Jones (2019) reminds us, is still evident, albeit in an implicit and diluted form. For instance, a textbook defines data as "raw facts that describe a particular phenomenon" (Haag and Cummings, 2009, 508), while the Royal Society (2012, 12) defines data as "numbers, characters, or images that designate an attribute of a phenomenon" (both definitions are cited in Jones [2019]). The constitutive element of what Jones (2019) calls a representational view, which in chapter 1 corresponds to maintaining a dichotomous separation between the physical/real and the virtual/digital, is the work that goes into crafting data. Underscoring the crafting of data as data from its inception to its later use undermines conceptions of data as naively representing a pregiven "reality" (Bowker 2014; Jones 2019).

Work, as feminist scholars remind us, is unevenly appreciated: the in/visibility of work is political. Invisible work is indispensable for organizational routines in the sense that it is "work that gets things back on track in the face of the unexpected and modifies action to accommodate unanticipated contingencies" (Star and Ruhleder 1996, 10). In the words of Jackson (2014, 223), invisible work is crucial when absent "systems seize up, and our sociotechnical worlds become stiff, arthritic, unworkable."[2]

A defining perspective in STS is shifting from, or at least supplementing, a focus on design (intentions, inscriptions, visions) to use (contingency, transformation, appropriation). It paves the way for an empirical research program on the appropriation of technologies throughout their life cycles (Williams and Pollock 2012). Consistent with such a perspective, repair, not only design and use, appropriation of technology is relevant (Jackson 2014). With digital technologies taking on infrastructural qualities, the role of repair expands. A working infrastructure, as Bowker and Star (2000) noted early, requires a lot of work. In a similar vein, Graham and Thrift (2007) urge us to consider

> all the processes of maintenance and repair that keep modern societies going. These processes can be likened to the social equivalent of the humble earthworm in their remorseless and necessary character—and in the way in which they have been neglected by nearly all commentators as somehow beneath their notice. Our intention is to bring these processes out into the light and to make them into the object of the systematic and sustained attention that they surely deserve to be, since they are the main means by which the constant decay of the world is held off. (1–2)

Vividly illustrating the role of repair and maintenance through the case of a city, they go on:

> Our laboratory will be the contemporary city, which hosts and is to a large extent defined by the myriad functions of maintenance and repair which themselves produce much of what might be regarded as the stuff of urban phenomenology. Think only of some of the familiar sounds of the city as an instance: from the sirens denoting accidents, to the noises of pneumatic drills denoting the constant upkeep of the roads, through the echoing clanks and hisses of the tyre

and clutch replacement workshop, denoting the constant work needed just to keep cars going.

The above general insights about the necessary role of forms of invisible work apply also in the context of data science (see, e.g., Fiske et al. 2019; Parmiggiani et al. 2021). Empirical studies of data science document the extensive and varied work implicated in data-driven approaches. There is, as Edwards et al. (2011) point out, data friction. This friction, and the efforts involved in removing or working around it, is precisely what risks being abstracted away in inflated expectations of data-driven approaches.

Tying back to the point of departure for this chapter: visions about data science regularly abstract from the considerable *work* involved in producing the data. Accounting for this work is a central concern for a research agenda on the role of data science in organizations. It has, as pointed out in a systematic review (Günther et al. 2017, 200), until now gone underresearched, as "future research needs to empirically examine how different actors within organizations work with big data in practice." What types of work, then, go into making data amendable for data-driven approaches (Kitchin 2014)?

The notions of "gathering" or "collecting" data are misleading, inasmuch as they promote the misconception that data speak for themselves. This downplays to the level of nonexistence the way data provenance—the methods, procedures, and technologies employed to generate the data— shapes data use and interpretation. As Gitelman (2013) notes, data "are always already 'cooked' and never entirely 'raw'" (3).

Not only collecting but also curating data involves efforts. Data quality involves maintaining procedures (see Leonelli 2014). Edwards (2010) examined the comprehensive data-gathering process informing climate change research and reports that measurement devices such as thermometers must be constantly calibrated to ensure the validity of their readings. In this context, maintaining calibration involves adhering to protocols that compare a given thermometer with a master device and systematically adjusting historic measurement values after discovering that a thermometer is uncalibrated. Similarly, in a study of a thirty-year effort to gather data to develop knowledge about HIV/AIDS, Ribes and Polk (2015) describe how maintaining subjects' commitment to contribute data—including various types of biological samples, interview data, and family hereditary data— over time involved updating subjects with relevant information regarding the progress of knowledge about the condition and conducting sustained persuasion campaigns lobbying for subjects' continued participation.

In the works of Orr (1996) and Suchman (1987), *repair* has focused on the situated and practical work that in Suchman's terms "must be contingent on the circumstantial and interactional particulars of actual situations" (185). Beyond keeping infrastructures going (Graham and Thrift 2007), repair has generative and productive connotations. Breakdowns are neither barriers nor catastrophic in a repair perspective. Rather, it is "precisely in moments of breakdown that we learn to see and engage our technologies in new and sometimes surprising ways," Jackson (2014, 230) points out, with more than a fleeting resonance with Heideggerian perspectives on technology (Ciborra and Hanseth 1998; Feenberg 2012). Similarly, in a more poetic language, Jackson (2014) elaborates on how repair

> occupies and constitutes an aftermath, growing at the margins, breakpoints, and interstices of complex sociotechnical systems as they creak, flex, and bend their way through time. It fills in the moment of hope and fear in which bridges from old worlds to new worlds are built, and the continuity of order, value, and meaning gets woven, one tenuous thread at a time. And it does all this quietly, humbly, and all the time. (223)

Central to visions about data science are the frictionless access to and the open-ended manipulation of data. This presupposes the repurposing of data gathered for one purpose to another, which begs the question "What counts as data—for whom, when, where, and why" (Leonelli et al. 2017, 195). Data, when collected, come with a preunderstanding of what made them data in the first place. Wylie (2017), drawing empirically on how archaeologists, much like geoscientists, work with data, points out that "as often as not the process of repurposing legacy data calls into question the very preunderstandings that made it possible to 'capture' these data in the first place" and refers to the profession that field archaeologists "have developed to continuously build and rebuild the scaffolding for evidential arguments that are understood to be provisional" (205).

THE CASE OF DATA MANAGERS IN NORTHOIL:
CRAFTING DATA INTO DATA

The theoretical perspective outlined above underscores the inherent presence and importance of infrastructural work (Bowker and Star 2000)—cleaning, repairing, articulation—in short, massaging data—to craft data. Highlighting the work to make data is not a novel insight. As made clear in the theoretical outline above, other scholars are making similar arguments. Accordingly, as demonstrated further below the infrastructural work implicated in the practices of knowing geodata is analytically not novel. However, it provides the necessary empirical detail to chapter 3's account of the siloed nature of geodata and, more importantly, the countertactics, -strategies, and -measures triggered.

To demonstrate infrastructural or data work, this chapter focuses on the work of *data managers*. Data managers are a group recruited from different disciplines and with varied professional experiences, including but not limited to production engineers, reservoir engineers, geologists, and information technology professionals. They also include administrative personnel who "have received on-the-job training," as one informant explained. They make a motley and heterogeneous group. Not all oil operators have formally designated data managers. They tend to exist at the larger ones. Still, even without designated data managers, the *tasks* detailed below need to be performed.

The data managers' work presented below is from NorthOil, one of the larger oil operators operating on the Norwegian continental shelf. NorthOil has different categories, or types, of data managers. Data managers are not organized into independent organizational units but are seamlessly embedded within other corporate units, such as production, drilling, and wells, reporting to national petroleum authorities or exploration. The case of the data managers presented here focus on the latter type tied with exploration, as this simultaneously acts as a backdrop for chapter 5, which empirically analyzes geological exploration and the work of explorationists (*letefolk* or *tolkere*, the group of mainly geologists and geophysicists involved in exploration; the name "explorationist" is their own vocabulary).

The presence of a data manager embodies the extent of the invisible work turning geodata into workable data for the explorationists. Had, as visions of data science suggest, data been available at the explorationists' fingertips, data managers would fill no meaningful role. As it happens, data managers make up about 10 percent of the manpower in the department of exploration at NorthOil. Explorationists and data managers work closely, practically seamlessly or symbiotically, together. For outsiders like researchers doing fieldwork, it takes a while to distinguish data managers from explorationists. Data managers are colocated with the explorationists in the same corridor, sharing meeting rooms and informal coffee areas. With office doors normally ajar, a data manager will pop into the office of one of the explorationists for a quick chat to clarify their request for data:

> "Do you have a minute?" one explorationist asks as he steps into the office of two PDMs [project data managers] busily working. "No," one of the PDMs responds, laughing. "Thanks," the second PDM chips in, "I didn't dare to say [no] myself!" Retreating, the explorationist offers to come by later, before the first PDM tells him he was only joking.

The legitimacy stems directly from the challenges of siloed geodata outlined in chapter 3. Precisely due to the many ways geodata are not ready-to-hand for geoscientists in exploration, there is a need for data managers. The geographically defined oil and gas field (or *asset*) also constitutes a strong organizational boundary, thus giving rise to siloed data. An oil field has a life span of decades and hence develops significant organizational autonomy. This autonomy has, until recently, included "different naming conventions" for wells, for parts of the reservoir, and for equipment and reporting standards for what and where to report. As one data manager explained, using the example of reporting well pressure data, some oil fields use "[a designated, formal] well report," while others employ "separate files" in the storage area, thus complicating the search for the data. Moreover, the conventions around types of pressure data vary, too, with "some reporting mud [pressure] data [while] others report formally quality-checked [QC'ed] data."

Generating new geodata, especially seismic, is expensive, hence explorationists must make the most use of historical data that already exist, data that are rarely ready to hand. Multitudes of data are available from the four decades of oil exploration and production at the Norwegian continental shelf, residing across different databases and systems and in a multitude of formats. Data need to be found, imported, quality assured, reworked, and reinterpreted to be applicable to current exploration pursuits. The work involves grappling with different databases, file systems, file formats, standards, and integration technologies; the work amounts to patching together an infrastructure. The SQL queries data managers employ may "fill a full page on the screen" and, even with computing power among the fastest money can buy, may "have response times in days rather than minutes," as one data manager explained.

In short, geodata are relative to geographical *location/field*, conventions of *naming* (varying across fields but also throughout the life cycle of a field), *time* of capture, residence in a variety of specialized *databases*, *method* (seismic processing algorithms, projection), type of *equipment* from service companies conducting the data collection, *formal* status (project, corporate, public data), different degrees of *quality* assurance, and the degree of *trust* in the professional competence of those involved. Compounding the challenges for explorationists of laying their hands on relevant geodata is the

Table 4.1
Overview of the project data managers' tools for navigating for information.

Type of data managers' tools	Number of systems	Example of functionality
Data management tools	3	Collect, find, edit, manage, and transfer data
Data integration tools	2	Specify workflows relative to given databases
Project data store	1	Exploration project database
Corporate data store	1	Project database, QC'ed
Team sites (Microsoft)	Numerous sites, 1 system	
National geobank, Diskos	1	Public database provided by Norwegian Petroleum Directorate

fact that a few years ago when NorthOil merged with another oil operator, itself the result of a prior acquisition of yet another oil operator, the geodata for the previously independent companies were incorporated into NorthOil's. "But not everything made it across!" one data manager informant admitted. Given this, the everyday concerns of explorationists, such as "Can I have all geodata for [a designated area]?" "Is this configuration of the well [equipment, completion method, possible later sidestep drilling forked off the primary well] the latest and updated one?," are nontrivial, thus generating the need for data managers.

At the same time, however, data managers are conscious of being in a potentially vulnerable position. They embody the nonheroic articulation work that often goes unappreciated by many, notably managers. Their value is thus something that needs constant attention when, to paraphrase Latour, for data managers every day is a working day. As one managerial representative of NorthOil explained, data managers are unlikely to vanish, but they clearly expected "significant automation" of presently manual tasks (such as the loading of data into the explorationists' tools) and "machine learning based flagging of quality [of data] issues." In other words, this manager expected, much in line with historic studies of automation (Autor 2015), that certain tasks presently done by the data managers will be delegated to automation and that the job of data managers will evolve to include new tasks.

The perceived precarious situation of data managers creates everyday challenges. Tasked with ensuring sound data governance procedures, data managers are to tidy up the data of exploration projects. While working, the explorationists pursue a multitude of possibilities (see chapter 5). One example is the tracing of seismic horizons—outlining an underground rock surface by selecting (*picking*) well data and then interpolating between well data points using seismic data. A seismic horizon represents a consequential, preliminary interpretation of the seismic. Opening up his exploration project workspace, one explorationist sighed, "You can find hundreds of versions of a seismic horizon in a project, and there is really no way of knowing which one to trust." The data managers are expected to, upon completion of an exploration project, QC the data. This, however, requires explorationists, already busy with the next project, to go through and sort the useful horizons from those less so. Data managers are to ensure compliance with this procedural part of the corporate quality system but have neither the standing nor the authority to enforce it. "We have to keep nagging them," one data manager complained, "to make the explorationists identify what horizon to QC." At certain predefined decision gates in the exploration process—for example, when deciding to drill a prospect—the data and interpretations underpinning the decision need to be QC'ed according to NorthOil-institutionalized policies. In what follows, two detailed examples of the work, hence value, of data managers are analyzed.

CRAFTING THE IDENTIFY OF DATA

After joining an exploration project's early deliberations, the explorationists' initial problem is to locate relevant existing data. If the project is in a so-called mature area with substantial previous activity, there is an abundance of data. To obtain what they need, the explorationists rely on the help of data managers.

The data managers understand the nature and purpose of the exploration projects and what data may be relevant to those projects; they know how to search across various databases and use a variety of tools (depending on where the data are and where the data go) to load data into the explorationists' interpretation tools, including various forms of quality control upon loading. Others work on organizing and visualizing data, for example, into maps in geographical information systems. The data managers also develop tools for explorationists, such as search tools for use across databases, file structures to browse well data that are accessible outside web browsers, and tools to keep track of the status of data at decision gates in the development of an exploration project.

Geodata relevant to oil exploration reside in numerous databases. Norwegian petroleum legislation requires seismic, well, and production data to be made open to the national Diskos database, accessible for private and public institutions for a modest fee (see chapter 3). In addition, geological interpretations and QC'ed data that are not required to be uploaded to Diskos are available in numerous NorthOil internal databases (see table 4.1). Exploration projects establish workspaces populated with intermediate, preliminary results and interpretations. Members of the Diskos consortium may, and regularly do, trade their internal, nonpublic data. The immediate problem is, accordingly, deciding which of the many variants and versions of a particular data set to base your work on.

When we joined the project, the explorationists were struggling with a typical problem in early-stage projects—namely, to gain an overview of all available geodata in the geographical area of interest. The underlying reason why this is nontrivial is basic: they have to make sure everyone is referring to the same entities. In other words, their problem is to fix the *identity* of wells, fields, production facilities, and seismic campaigns. For instance, the naming of an exploration project varies over time. Initially, an area receives a designation local to the exploration project. The modest number of projects (compared to the total) deemed commercially interesting (see chapter 5) are equipped with a corporate name. As part of obtaining the formal granting to develop the field by the National Petroleum Directorate, all fields are given official names, typically drawn from Norse mythology (e.g., Brage, Valhall, Heidrun) or famous Norwegians (e.g., Ivar Aasen, Johan Sverdrup). In addition, naming conventions vary across service companies (for drilling, seismic) and fields and assets (historically, there were several dozen local conventions in NorthOil).

As a consequence, data managers cannot assume that the same well will have the same identifier qua name across different databases. The identifier (ID) of the relevant exploration data has to be *crafted*. A well log can have a different ID in Diskos than in NorthOil's internal database. To identify the same data set across databases, data managers must have intimate knowledge of the IDs across systems, as illustrated below.

One day, we sat in on a discussion that a group of data managers were having with one of their several consultants hired to do some of the grunt work of data managers. In addition to NorthOil-employed data managers, a number of consultants are temporarily hired to supplement such work. They work closely together, and the intended division of labor is for the consultants to perform the routine, less complicated tasks. The consultant was diligently working on ID mapping, constructing a list of synonymous names (*harmonization*) that are used in the different databases but that all refer to the same physical entity. Concretely, they discussed the harmonization of IDs for seismic files in Diskos with the internal workspace databases with (also preliminary) interpretations. The rationale for harmonizing internal IDs with Diskos IDs is to guarantee that all Diskos data available to NorthOil are made available for interpretation in the explorationists' current project workspaces. To account for his efforts, the consultant went through a spreadsheet he had created containing the workspace IDs and Diskos IDs, as well as information such as "not Diskos" or "not workspace database." The consultant, however, had no in-depth knowledge about the seismic files. His ID mapping was based on relatively superficial cues. He simply compared the IDs in the explorationists' project workspace with those in Diskos, estimating that he had "gotten 95 percent of the mismatches" and resolved these. The efforts of doing so were modest, as the heuristic was simple. This was not so for the remaining 5 percent, however. Is it worth the effort, the data managers pondered, to have a crack at sorting out these nontrivial cases? They discussed. Perfection is never attainable. It is invariably the more pragmatic concern of "good enough" that prevails. For the data in question, there was a certain demand from the explorationists. Accordingly, they agreed that the data managers, with the domain knowledge the consultant lacked, should carefully go over the remaining 5 percent. The data managers employ a number of heuristics and tactics, drawing on knowledge about prior exploration projects and local conventions for storage and naming as well as clues provided by the metadata available for the data set. They "look for when the file was created" as a clue to the date of the data in order to identify time-variant names. They "analyze the file endings" (akin to assuming that .pdf is a postscript file and .docx a Word document) as a proxy to determine the identity of the service company that produced the seismic and, therefore, what method and measurement devices generated the data.

Over coffee after the meeting, one data manager elaborated on the challenges just addressed. A key source of difficulties was that when NorthOil imported the exploration workspace data from the company they had merged with some years earlier, they maintained the legacy IDs without harmonization. This gives credibility to those concerned about the significant invisible costs (work) involved in large-scale corporate mergers and acquisitions

(Vieru et al. 2014). Moreover, when working on an exploration project, the official naming conventions often are "not fine-grained enough," prompting local names to be generated. The national authorities' official naming only applies to a complete oil field, not its constitutive wells and equipment.

On a different occasion, we discussed the efforts of data managers to determine not only the identity of data but its ownership. Seismic surveys are usually commissioned by an oil operator that is the "operator" (leader) of a *license* (a consortium of oil operators, each with a specified percentage of the interests in the license).[3] Hence, seismic surveys are most often owned by the consortium that made up the license at the time of the seismic survey. However, the ownership of a license consortium may, and regularly does, get traded, and hence changes, over time. The issue facing our data manager was that he had to produce a *merge* survey (a seismic survey that merges several existing surveys into a new one) as part of a new exploration license. The question, tied to potential lawsuits with significant economic risk, is whether the partners in the new license held legal access rights to the existing seismic surveys. The data manager explained his tactics. The coding of information built into the naming convention provides a starting point:

> The two first out of three letters [in the file name] are short names for the companies. Then there are the numbers, which is the year the survey was shot. Ninety-eight is 1998, 04 is 2004. And then there are three numbers at the end. These numbers indicate what kind of seismic it is. Is it the usual license seismic, or a site survey, that is, seismic shot as they are drilling a well.

The national geobank, Diskos, has an overview of all surveys that contain the name of the service company conducting the seismic survey. The surveys are conducted by a service company but owned by a license. In order to determine the ownership of the survey, the data manager must dig up old reports from the survey or its subsequent processing. What we must do, our data manager explained, "is to go through, survey for survey, to see what is registered at Diskos and on our [NorthOil] internal databases. If we are lucky, it will state that the survey was done in a particular license. But sometimes it does not say and we do not know. And when we do not know, we must locate the names of people from [NorthOil] to see if they remember."

CRAFTING OVERVIEWS OF DATA

A near-chronic state of affairs when grappling with digital oil data, after several decades of the massive generation of data as part of activities, is being overwhelmed by data. Similar to other forms of so-called knowledge work (e.g., physicians' work at hospitals; see Ellingsen and Monteiro 2003), explorationists feel "drowned" in data. As a consequence, they are uncertain as to whether they have succeeded in identifying all relevant data for their project in question. A vital rationale for data managers is their ability to round up all relevant data for explorationists. As pointed out by one data manager,

> There is a problem that a lot of data exist without being visible for the explorationists. Hence, it does not exist for them. The data is there in the databases but is not shown. If no one checks, if it is not made visible, the explorationists will miss out on potentially relevant information.

In response, some of the data managers have developed a geographical information system tool to show the existence of data using Python scripts to crawl the NorthOil corporate databases for seismic, well, and specialized data (petrochemical, geophysical), as well as license data. Similarly, the data managers draw on a well spider tool to search across projects. By default, data managers do not, for reasons of confidentiality, have access to all the data. The well spider tool, however, "is not widely used," one data manager acknowledged, as it is perceived as "too complicated." Instead, the data managers resort to consulting with a senior data manager to learn who is responsible for data in the area of interest and approach them. Data managers, to compensate for their distributed and embedded existence across corporate units and projects, develop their own "marked place" for sharing hints, practical advice, resources, and tools.

The above example addresses crafting overviews across different *geographical* areas. Other types of overviews, too, are regularly crafted by the data managers. One example is keeping overviews across *time* by ensuring that the most recent data set is used. This requires oversight into operations beyond the local ones in the project the explorationist is working on. "Sometimes the explorationists are frustrated that the well data [they require] is not available for them in their project," but as one data manager went on to explain, this "is because the data changes in the underlying database, for instance when a drilling operator changes a well log. This then needs to be reloaded into the project database [before the explorationists can access it]."

In a similar vein, the exploration manager responsible for exploration across multiple projects struggles with keeping track of which versions of geological interpretations are actually used to underpin particular geological interpretations. This is a particularly pertinent problem when it comes to documenting the data used to make decisions at the formal NorthOil-institutionalized decision gates that regulate the selection of which oil prospects in a portfolio are prioritized and why (see chapter 5). As a remedy, the data managers developed a dashboard tool that allowed explorations' managers to monitor the status of data in projects they were responsible for, as one data manager explained: "Now they [exploration managers] can see what seismic and wells [data] they have in their [portfolio] and what the status of the interpretations are. . . . They can see what data has gone into a decision gate, and if it has been quality assured."

A final illustration of the relevance and importance of overviews across time-space is keeping track of work across projects. "I know that the time-depth curve I have is wrong," sighed one explorationist as he crashed into the office of one of the data managers. The *time-depth curve* he referred to is a crucial yet challenging part of being able to combine the two data sets most central to exploration—namely, seismic and well logs. The problem with combining them is that they are, literally, noncommensurable. Commensurability needs to be crafted (Espeland and Stevens 1998). Seismic images are generated by measuring the *time* it takes for acoustic sound waves to be reflected from the intersections of different subsurface layers. Different layers of rock yield different wave reflections and, as the core of the time-depth-curve problem, depth is inferred from the time delay of acoustic waves. But because the speed of the waves depends on the types of rocks in the layers, the time-to-depth conversion (time-depth curve) is a nonlinear mathematical function calibrated with well log data that measure the actual depth along the well bore. The conversion is known as *well-time*, the correlation between the time of a seismic wave reflection to the depth of a well log. The explorationist in the example above is approaching the data manager to find the well-time conversion conducted by another project, as "they have their well-time and we have ours. I want their well-time because they are the ones who have worked on it most recently," thus assuming that the newer data (and conversion) are more accurate than the older data.

CONCLUSION

An emphasis on infrastructural work (invisible work, articulation work, or repair work) is, with more than a nod to Braverman (1974), an antidote against automation as "the politics of knowledge repair [will try to] deskill repairers" (Graham and Thrift 2007, 18). Consistent with practice-oriented studies' demonstration of the irreducible entanglement of users in working digital technologies, data are neither given nor captured but crafted. Edwards et al. (2011) argue that simple technological solutions to what they call ontological incompatibilities are unlikely, as "we have not yet developed a cadre of metadata workers who could effectively address the issues, and we have not yet fully faced the implication of the basic infrastructural problem of maintenance" (10).

If the kind of infrastructural work currently performed by NorthOil's data managers is unlikely to be automated, the boundary is likely to shift in the sense that selected tasks of their work get automated (Autor 2015). The pursuit of new deep learning–based algorithms for working with dirty and noisy data sets is a fiercely active agenda with considerable recent progress (see, e.g., Algan and Ulusoy 2020).

In the case of oil exploration, dirty and noisy IoT data prevail (see chapter 3). New algorithmic methods for cleaning or repairing data are making headway (e.g., removing outliers). However, the principal resource of the data managers stems from their *domain* knowledge of previous exploration projects and their ability to navigate the many databases and digital tools containing potentially relevant digital oil data. In our case the hired consultants are, accordingly, significantly more vulnerable to the threat of automation than the data managers. Data managers, then, illustrate the crucial importance of grasping the domain (context) of data (see also Ribes 2019; van den Broek et al. 2021). Data managers thrive on their embedded relationship with explorationists and their projects. Viewing data managers as merely mediating data to the explorationists misses how they need to collectively perform. The cumulatively acquired familiarity data managers have with the metadata (e.g., dates, file types, service companies' equipment) enables data managers to craft the data for data-driven exploration. Being sensitive to the incomplete and cooperative characteristics of repair enables us to "redirect our gaze from moments of production to moments of sustainability and the myriad forms of activity by which the shape, standing, and meaning of objects in the world is produced and sustained" (Jackson 2014, 234).

Empirically underpinned analyses of how, where, and when domain knowledge meshes with data-driven approaches like the above account of data managers' practices are rare (exceptions include van den Broek et al. 2021). One example is Passi and Jackson's (2018) analysis of a commercial company's efforts to predict *churns*—that is, currently active customers who are likely to cancel paid services in the future. In internal company discussions, there was an ongoing negotiation about the trade-offs at play in the configuration of the parameters driving the optimization yielding churn predictions that evolved around domain/business insights versus data science methods criteria:

> As a pragmatist, what I am looking [for] are things that are highly sensitive, and their sensitivity is more important to me than their accuracy. . . . If you can ensure me that of [say] the 2700 [customers] we touch every month, all 500 of those potential churns are in that, that's gold for me. . . . If you could tell me [to] only worry about touching 1000 customers, and all 500 are in it, that'd be even better. But . . . let's start with [making] sure that all the people I need to touch are in that population, and make maximum value out of that. . . . It is about what outcomes I am trying to optimize to begin with, and then what outcomes am I trying to solve for and optimize after. (136)

Data scientist professionals are in high demand. Many of the large technology companies, including but certainly not limited to Amazon, Google, and Facebook, are proactively recruiting professionals with a strong profile in data science and machine learning. Universities scramble to establish teaching programs to meet the demand for data scientists. A crucial question is exactly what kind of competence is required to meet the varied demand for data scientists in the future. Our analysis of the role of data managers provides guidance.

A purely inductive, data-driven emphasis on data science suggests the need for data scientists with deep, specialized competence in the underlying statistics of data-driven and machine learning–based methods. Some organizations develop centralized data analytics centers to govern and control tracking, collecting, managing, processing, and analyzing big data for decisions (Günther et al. 2017). Undeniably crucial for some, the analysis in this chapter strongly suggests a different competence profile that combines the basic skills of data-driven methods, many of which are available in commodified tools, with a deep, relevant knowledge of the domain (cf. Davenport 2014). Only then may data friction and multiplicity be appreciated, without which data science methods in business organizations will be limited. As noted by Constantiou and Kallinikos (2015), data science "owes much of its distinctiveness to the mechanisms by which it is generated and the messy or trivial everydayness these mechanisms help install at the heart of the processes of data generation and use" (46).

5 UNCERTAINTY

written with Marius Mikalsen

The practical problem par excellence: what to do next?
—Garfinkel (1967)

Acting, always, is uncertain (Callon et al. 2011), with different types of uncertainties being conceptualized as unknown, epistemic, Knightian, or black swans (Beunza and Garud 2007; Faulkner et al. 2017; Le Masson et al. 2019). The rise of calculative methods (i.e., methods of quantification) to the "taming" of chance was historically an institutional and organizational response to this uncertainty (Hacking 1990). Unruly qualitative chance gradually got tamed, or, more accurately, taming was attempted (Beck 1992), by quantified conceptualizations of risk (Douglas and Wildavsky 1983; Porter 1996).

This chapter deals with the actions of the explorationists, the community of mostly geologists and geophysicists feeding off the data unearthed by the data managers analyzed in the preceding chapter. The explorationists represent a vivid example of acting under uncertainty, an uncertainty grounded to a large extent in vast yet underspecified and uncertain data. The explorationists' actions are not only uncertain but also consequential. The phase of exploring for hydrocarbons represents a vital, strategic investment for global upstream oil operators. Effectiveness in exploring is thus decisive for long-term competitiveness. Yet, as one explorationist explained, they are used to "a hit-rate [the rate of hitting commercially viable hydrocarbon reservoirs among the total of drilled wells] of about 5%."

The lure of data-driven data science is predictions (Agrawal et al. 2018), when we take the classification of patterns offered by deep learning with convolution networks as a form of predictions too (Marcus 2018). Data-driven methods from data science thus come with a potential, if not promise, of guiding, supporting, or indeed automating acting under uncertainty.[1] The naive belief of a purely inductive *no theory* approach is exactly that, naive. Still, the ongoing discourse on the motley set of techniques collectively known as data science or artificial intelligence (AI) comes with intriguing examples of interesting applications as well as stark criticism (see chapter 4). To explain the position adopted here, it is illuminating to briefly review the previous round of hyped-up AI some of us found ourselves in the midst of a couple of decades ago.

AI has always invoked a powerful imagery. Machines capable of human reasoning are captivating. AI emerged in the 1960s–1970s to grow into something of a hype during the 1980s–1990s. After a stretch of less attention to AI, we are presently experiencing a renewed interest in it. A comprehensive review of the history of AI, comprising its numerous variants in methods, technology, and application, is significantly beyond the scope of this outline. Here it suffices to sketch two distinct approaches, the relevance of which is the difference in their role of data in the knowing process.

The first approach to AI, pioneered in the 1960s–1970s and maturing over the following couple of decades, aimed at mimicking human cognition (Newell and Simon 1976). Known as symbolic AI, the approach was geared toward formalizing domain knowledge and reasoning through rules and deductions in a knowledge-based system, an expert system, or a decision support system. Information systems based on symbolic AI approaches met with criticism. Conceptually, scholars argued that human reasoning intrinsically escapes what is possible via symbolic AI (e.g., Suchman's [1993] argument about situated action or Dreyfus and Dreyfus's [2000] argument about embodied action). Empirically, symbolic AI–based systems largely failed, despite significant and prolonged investment, to be institutionalized into organizational decision-making practices; hence, the challenge remained "to integrate the decision making of these systems among functional, planning, business, and global units of organizations" (Wong and Monaco 1995, 148).

The second approach to AI is presently unfolding. It aims at mimicking human biological rather than (directly) cognitive processes. Neural nets, especially in the form of deep learning,[2] simulate neurobiological processes of the brain. Born during the first wave of AI, the recent reemergence of data-driven subsymbolic AI has gained traction. Data-driven, subsymbolic AI makes up the backbone of current data science methods, especially following the booming attention deep convolution networks received after the pioneering research in computer vision around 2012 (Krizhevsky et al. 2012). The reactions to this second wave of AI are markedly polarized. Some argue for far-reaching, bordering on revolutionary, implications for human reasoning and decision-making because "we know the opportunities and impact of big data will be substantial" (Davenport 2014, 28; cf. McAfee et al. 2012). Others insist that the fundamental problems with the first wave of AI prevail—that is, nothing essentially changes with subsymbolic AI (Dreyfus and Dreyfus 2000; Suchman 1993).

Rather than an a priori, bordering on ideological, stance on the limits of AI, this book advocates an empirically open, explorative approach. Automation, with regular setbacks and challenges, is shifting its focus to currently targeting the heartland of the qualitative—that is, sensemaking, interpretation, and common sense in ways that cannot be brushed away (Autor 2015; Shrestha et al. 2019; von Krogh 2018). Data-driven data science "works" in ways old AI never did, which is simultaneously unsettling and intellectually intriguing. Rather, the key is to critically analyze the social, material, and institutional circumstances underpinning actual, not merely potential, data science practices. As Lyytinen and Grover (2017, 226) phrase it, there is a "need to articulate more refined technology-imbued theories of data origination, use, management and control." This is consistent with observations in the literature on the dire paucity of empirically grounded studies of how data science methods influence decision-making and knowing practices in organizations (Günther et al. 2017).

THEORETICAL FRAMING: ORGANIZATIONAL DECISION-MAKING

The uptake of purely inductive, data-driven approaches into organizational work practices is slow, if it occurs at all (Günther et al. 2017). Chapter 4 accounted for one important reason for this—namely, the underapprecia-tion of the infrastructural (repair, articulation, invisible) work to craft data.

The present chapter expands into the necessary conditions and mecha-nisms for the actual uptake of data-driven work practices. It zooms in on how data underpin organizational decisions. More specifically, the focus is on how *consequential* decisions are forged. In contrast, the large class of recommender systems mostly address consumer choices (Bobadilla et al. 2013)—for example, what additional Amazon book to consider after you have already purchased one. The significant, and in the context of this book, relevant, implication of a focus on consequential, organizational decision-making, not relatively inconsequential consumer choices, is the deliberations and negotiations regarding the data underpinning data-"driven" decision-making.[3]

Data-driven approaches, understood literally, downplay to the level of nonexistence the insights from institutional theorists on the relationship between data and decisions. As March (1994), among others, spent a lifetime documenting, organizational decision-making does not in any way flow from data. Organizational decisions, institutionalist perspectives teach us, are not data driven in any meaningful sense of the expression. The idealized information-processing model of rational decision-making, in which decisions are unilaterally given by a weighted sum of the data (information)—implicitly if not explicitly assumed in data-driven approaches in data science—lacks empirical, organizational grounding. For instance, instead of data driving decisions, the relationship in organizational decision-making may be the exact opposite: decisions are made first, with data employed for the legiti-mization of already made decisions (see Feldman and March 1981). Key concerns in practice, left unaccounted for in a purely data-driven approach, evolve around issues of the legitimacy, accountability, and credibility of data (Kitchin 2014).

The theme of this chapter is to analyze how data—in different connota-tions of the notion—are marshaled into credible evidence in institutionalized decision-making processes. At the core is an analysis of how actual data-driven decisions are institutionally forged in the defining condition of digi-tal oil—namely, that despite its vast, heterogeneous nature, data chronically underdetermine the phenomenon under scrutiny. With inherent epistemic uncertainty, how do practitioners cope?

An illuminating empirical illustration is the data-driven crafting of meteorologists' operational predictions examined by Fine (2007). He demonstrates how meteorologists develop weather forecasts based on incomplete data and models. He highlights the collective practices whereby meteorologists discuss and challenge each other's forecasts, thus creating a peer-based system of accountability. Meteorologists are painfully aware of the incompleteness of their epistemic objects—the weather patterns over the next couple of days—but this very uncertainty makes the validation and justification of predictions all the more organizationally real and accountable: "That organizations are committed to accountability makes decisions social and political as well as scientific" (Fine 2007, 193). Moreover, procedures of verification are collective exercises that conjure credibility (Power 1997), which means that "verification is an organizational practice" (Fine 2007, 194). Similarly, Pollock and Williams (2016), in a study of the justification of Gartner's predictions of technology trends, observed quasi-scientific procedures of peer reviewing, protocols for what constitutes "evidence" that yield a collective regime of accountability; more precisely, when grappling with uncertain, partial knowledge, it is crucial to legitimize and justify predictions to make them credible and not mere guesswork (see also Tuertscher et al. 2014).

The focus in this chapter is how practitioners grapple with inherently inconsistent, incomplete, and uncertain data when organizationally legitimizing decision-making and action taking. As Pollock and Williams (2016) point out, a lack of hard evidence does not imply that decisions and predictions are mere speculation; they are subject to regimes of collective accountability that strive to make the most of the evidence at hand (Tuertscher et al. 2014; Fine 2007). If hard evidence is difficult to come by, are we left with mere speculation? Hardly, as pragmatism is all about. Rather than a quest for truth and absolute certainty, pragmatism accepts fallibility as a founding principle and analyzes how our beliefs are justified in the absence of hard evidence. For pragmatism, "standards and tests of validity are found in the consequences of overt activity, not in what is fixed prior to it and independently of it" (Dewey 1930, 72).

As Wylie (2017, 213) notes using archaeological interpretations as an example, it is crucial "'to honor ambiguity' rather than smoothing, cleaning, and otherwise suppressing the uncertainty of the jointly descriptive and interpretative claims that become the basis of subsequent reasoning with archaeological data." Known from pragmatism (Dewey 1930, 72), abduction is recognized as a practical problem-solving process seeking new associations, neither inductive nor deductive, to navigate inherent uncertainty and ambiguity. Organization scholars have long appreciated the role of abduction. For instance, Dunne and Dougherty (2016, 132) demonstrate how scientists in the biopharmaceutical industry apply abductive reasoning as a "deliberate and methodological" social process to "navigate in the labyrinth" of drug innovation.

Notwithstanding abduction's practical, problem-solving focus, abduction involves shortcutting searching for "innumerable possible hypotheses all accounting for the data at hand" (W. M. Brown 1983, 401). Charles Saunders Peirce underscored the principle of parsimony regulating abduction, a principle corresponding closely with what neoinstitutional scholars on organizational decision-making have identified as satisficing (March 1994).

The parsimony or satisficing principle regulating abduction sets boundaries (resource, time) for an otherwise open-ended process. Good-enough solutions ensure arriving at a decision within set limits. This, however, assumes that you know what you are looking for. In many situations you are, as then secretary of defense Donald Rumsfeld famously commented, not looking for "known unknowns" but rather "unknown unknowns" (Faulkner et al. 2017; see also March's [1991] distinction between exploration and exploitation). Similarly, Stark (2011) points out in the context of business organizations how "many firms literally do not know what products they will be producing in the not so distant future" (21). There is, in other words, a multiplicity of interpretations that abduction cannot eliminate.

Multiplicity is not only irreducible but also productive, indeed necessary, in situations of epistemic uncertainty. With "unknown unknowns," multiplicity is productive. It provides the necessary preconditions for revisiting, challenging, and, possibly, radically changing the interpretation of data, a characteristic of the discipline of geology (Bond 2015).

The operators in NorthOil, resonating with accounts of pragmatic action (Dewey 1930), operate under inherent uncertainty. They strive not for perfection in their models but for the practically useful. They are painfully aware of the contingent, temporal nature of the knowledge underpinning their daily operations. They know from personal experience that "the primary value of models is heuristic: models are representations, useful for guiding further study but not susceptible to proof" (Oreskes et al. 1994).

The Case of Oil Exploration: Grappling with Inherently Uncertain Data

Oil reserves, commercially viable reservoirs of hydrocarbons, are the life-blood of upstream oil operators in the sense that they secure long-term economic performance once the presently producing, revenue-generating fields gradually get depleted. The quest to locate and identify—explore— new oil fields is thus of crucial strategic importance. It typically represents 10–20 percent of total investments for an upstream oil operator. For national petroleum authorities, too, the estimated remaining oil reserves of the oil operators are vital and thus are part of the annual reporting scheme.

However, commercially viable oil reservoirs are extremely difficult to locate. Oil exploration is a decisively knowledge-intensive, data-driven endeavor. The term oil "reserves" is potentially misleading, as it suggests a level of certainty akin to having bank deposits. In contrast, oil reserves need to be recognized as, in their vocabulary, a *prospect*, a candidate for a commercially viable oil reservoir with varying and evolving estimated chances of actually containing hydrocarbons. A prospect, like financial instruments, comes with different and dynamically changing opportunities and risk. Prospects are, again as in finance, managed by oil operators as portfolios of investments. A portfolio of prospects is shepherded through a staged or gated process, which at NorthOil is known as a funnel model (figure 5.1). Using the example from the North Sea, one explorationist explained the funnel model as starting by "work[ing] on very large scales, [from] the entire central North Sea, down to [squares where sides are] three to ten kilometers in the prospects." As they narrow down potential reservoirs, a basin model, which is a large-scale but coarse geological account (cf. the narrative form of geological knowledge indicated above) of the area that explains how hydrocarbons were generated, migrated, and captured in geological time, must be established. For

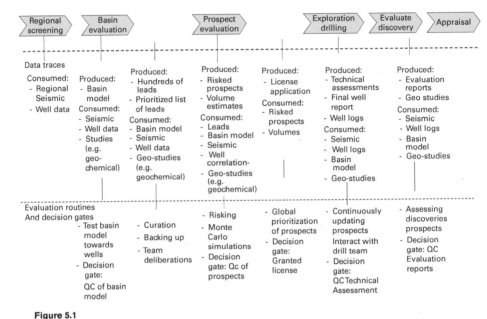

Figure 5.1

An overview of the funnel model for the life cycle of prospects together with their underpinning data and associated evaluation routines.

Source: Artwork produced by Marius Mikalsen.

instance, most basins in the North Sea consist of sediments of plankton captured in sandstone from the Jurassic geological period, when they formed the seafloor. The next step is to locate prospects that are candidates for drilling. These prospects are quality controlled, potentially approved as drilling candidates, and ranked among the full portfolio of NorthOil's prospects. Pending company prioritization, license applications are made to the Norwegian government. Following a successful license application, either an exploration well is drilled or the license is dropped (e.g., if the risk is too high or the predicted volumes too small). If drilled, the results are evaluated. If the discovery is significant, it is then appraised, which may result in the drilling of additional *delineation* wells to determine the size of the oil field more accurately before developing it for commercial production.

The work of elaborating a prospect—that is, investing in improving the estimates for an oil find, is organized into project groups consisting of seven to ten explorationists. As explained in chapter 4, the self-designated term *explorationist* covers those in several subsurface disciplines, including reservoir engineers, petroleum engineers, and petrochemical engineers, but is dominated by geologists and geophysicists. Obviously a simplification yet still helpful when getting into the work of explorationists, geologists and geophysicists belong to different epistemic communities (Knorr Cetina 1999): they differ in their vocabulary, methods, and perspectives (Mol 2003). Geology is qualitative in its orientation (Frodeman 1995). In a hermeneutical process, geology strives toward working out the geological processes that resulted in the present. Like archaeology (Wylie 2017), geological knowledge of these processes is essentially narrative in form. In contrast, geophysics is oriented toward capturing the present based on quantified measures of Internet of Things (IoT) data. In working with prospects for NorthOil, these two orientations among the explorationists are interleaved and contested. The central problem in oil exploration, starting from measured observations of the geophysical properties of the geological formations evident today, is to tie these to an inferred narrative account of the rich geological processes (erosion, sedimentation, tectonic plate movement, diagenetic processes, faults, and so on) that could have yielded the current situation. As described in chapter 4, the explorationists are colocated with data managers. This chapter, however, focuses on the work of the explorationists, not the data managers.

The work of explorationists is to refine the prospects by comparing them with available data and making models and simulations as well as collective deliberations among the community of explorationists. The life span of prospects (when not terminated) in the funnel process is several years. Verifying the predictions—the proof of the pudding is in the eating—by actually drilling an oil well lies many years ahead, if at all. Given the significant cost of drilling, about USD $100 million per well on the Norwegian continental shelf, drilling is done only after the predications are extensively refined. The everyday work of explorationists is, accordingly, devoted to elaborating, corroborating, refining, and challenging prospects that are so much more than guesswork yet fall significantly short of hard evidence (Pollock and Williams 2016). The focus of the empirical account that follows is on how explorationists *do* predictions.

In chapter 3, the different types of geodata were outlined and some of their data-quality issues characterized. Chapter 4 went on to detail how data managers did important infrastructural work to help the explorationists navigate some of the hurdles of geodata. Despite this, however, there are imperfections with the digital oil data of explorationists that shape their everyday practices. Notably, the data of explorationists are inherently incomplete, inaccurate, and inconsistent, giving rise to pragmatically regulated sensemaking. In what follows, three sets of explorationists' work practices are highlighted. Their separation is for analytical purposes only, as they interleave empirically.

CONTINUITY: ACCUMULATING "EVIDENCE"

The daily hum of explorationists' work is dominated by accumulating "evidence" for a prospect. The focus is on backing up a given prospect. In mature (i.e., brownfield) areas like the one empirically analyzed here, explorationists will typically bootstrap this by starting from what they call a *proven play*. A geological trap constitutes the necessary but not sufficient conditions for an oil reservoir and requires the three ingredients of a source (e.g., sandstone with plankton sediments), a migration path (e.g., a geological fault), and a seal (i.e., nonpermeable rock; see figure 2.1). The three elements of a geological trap, however, only set the general structural requirements for an oil reservoir. Globally, they come in numerous empirical instantiations (*plays*). A proven play, in the present context, is a particular instantiation of a geological trap in the area of the North Sea that has previously resulted in an oil discovery. In the words of one explorationist: "In a mature area, such as the North Sea, we know that there are several plays already that have been proven through drilling and discoveries. [So] you have the same concept . . . but [apply it to] new data to create new opportunities." Quite reasonably, then, explorationists start by looking for the same proven plays in other neighboring areas; they start by looking for known unknowns (Faulkner et al. 2017).

Equipped with a proven play, the explorationists focus on gaps, areas where no discoveries have been made yet but that are in the vicinity of existing oil reservoirs. These gaps also come with incomplete data at hand because the crucially important well log data (see chapter 3) by necessity are only available where drilling has already occurred. In a clear example of the pragmatic sentiment of making the most of what you have, the explorationists "stretch" their data by extrapolating it to fill gaps. As one explorationist described:

> Obviously, if you have three wells, they're going to tell you a lot about the vertical [well paths]. So, you have at least the understanding of the vertical sense of the layers and you can build your sedimentological understanding. . . . You have three wells and . . . you try to interpolate between those wells with your information and then you try to extrapolate away from those wells into areas that are further away. And then with the help of the seismic, you try to calibrate and use the seismic to help you, and then come up with some sort of feeling about whether, you know, how much reservoir you've actually captured with the data you have? (Monteiro et al. 2012, 99)

One tactic is to analyze gaps to determine if it is geologically possible that hydrocarbons may have migrated from an existing oil reservoir with a proven play to an area nearby. As an exploration team leader explained, "In this area [pointing to his screen], we knew that in the southern [name of the basin], which in this case is 250 kilometers north-south, a lot of hydrocarbons have been generated. So, how far east can those hydrocarbons migrate?"

As will surprise no infrastructure scholar, particular attention is paid to areas near existing production installations. Installed pipelines and processing and transportation facilities represent significant sunk investments for oil operators. They exert inertia in the sense that supplementary oil fields that can tap into the existing production installation bring down marginal costs radically compared to building it from scratch. One explorationist tasked with ranking prospects in the portfolio commented that, despite its modest size, one particular prospect was attractive due to its vicinity to existing infrastructure: "This [prospect] is very small, but it is close to [production, processing, and transport] infrastructure, so that is our winner."

Models are central vehicles for capturing and articulating the insights of the explorationists. Models come in radically different levels of elaboration and sophistication, ranging from crude outlines of a basin model to fine-grained, finite-element three-dimensional (3D) simulation models equipped with permeability and porosity measures (see figure 5.2).

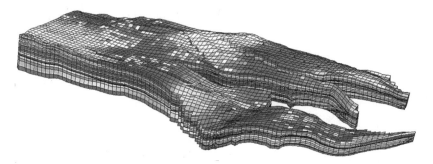

Figure 5.2
Simulation-based geological modeling of an oil reservoir.
Source: Reproduced by permission from the Norne open data set.

As part of refining their predictions, accumulating more support for a prospect, models are elaborated on demand. One member of the exploration project offered a good illustration of how he worked to incorporate historic data to back up his model. "So, we have a nice trap," he explained sitting in front of his screen, "but the question is whether there is any migration of petroleum toward that trap." His task was to analyze historic data of nearby wells to see if they meshed with his evolving model. Concretely, he was checking to see if the data pertaining to the age of the source rock supported the hypothesis of a possible migration. He explained:

> I use temperature and vitrinite reflectance of the source to get an idea of the maturity, and then I combine that data with information from the wells.[4] Hence, I build a model. But first I need to see that my model matches the other wells that I have in the area. If it does not match, then I need to reformulate my model until I get a good match to the wells. Here, the good thing is that we have a lot of [existing] wells. So, the uncertainty is relatively low. Then, I use the basin model to extrapolate what we know—based on the areas that we have drilled—to those areas where we have not drilled [the targets]. I try to use all the wells in the area, but it depends on the type of data, and the type of measurements that were done.

Even if the explorationists, as most scientists (Oreskes et al. 1994), are acutely aware that "all our models are wrong," as they jokingly put it, "by definition," they simultaneously acknowledge an emotional and psychological bond that develops after months, even years, of working on modeling a prospect: "It is the psychology behind it because when you work on something for a long time, you begin to think this is great, we have to drill it," and "Some people get really personal, and if new data goes against it, they try to go avoid the data."

An important way the explorationists improve their predictions—that is, the credibility of their prospects, is to iron out the considerable level of inconsistencies found in digital oil data. A key tactic is to link—visually and manually (see figure 5.3)—between the two principal types of data, well logs (detailed but narrow) and seismic (crude but covering large areas).

Even when you do have the data, quality is an issue. It is highly dependent on the purpose of its collection. For instance, a couple of decades ago well logging focused on the deep levels because they corresponded to the geological era of identified interest, Jurassic. More recently, explorationists have become interested in earlier geological eras with shallower stratigraphic layers of well logs, "but when we go back in time, the shallow levels were not logged properly [i.e., data quality is poor], only the deep levels."

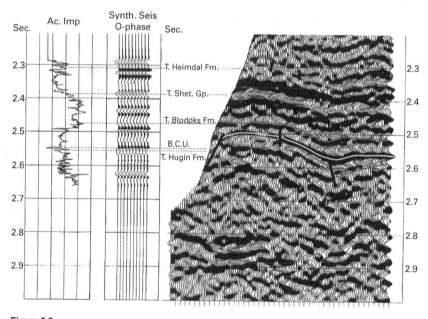

Figure 5.3

Linking well log data (detailed, narrow) with seismic data (crude, broad) by visually linking different rock combinations likely to correspond to each other in the two data sets.

Source: Reproduced by permission from the Volve open data set.

Although explorationists demonstrate a robust tolerance to inconsistencies by ignoring or downplaying them, some types of inconsistencies need addressing. One source of inconsistency stems from issues with old versus new data. One explorationist, sitting in front of his screen, was struggling to use a petrophysical analysis tool filled with well-log data. The old well data he had available was not compatible with his prospect. Was he or the data wrong? he wondered. The well data in question were dated. They were generated when the well was drilled back in the 1970s. As he explained, the knowledge that injecting mud into the borehole while drilling influences the temperature readings in the well "was learned only in the 1980s." The old well data measurements were, accordingly, not to be trusted. Instead of outright contradicting his prospect, he assigned a mark signifying that the well data were of low quality to indicate their lack of relevance.

An important task in determining a prospect's credibility is to see whether it fits several types of data in what effectively corresponds to a form of triangulation. So-called well tie-ins are a particularly important way in which this triangulation operates (see chapter 4). Digital interpretation tools are used to determine the relationship between boundaries in the well logs and seismic reflections, consequently producing a relationship between the well logs (measured by depth) and the seismic reflections (measured in time). A well tie-in is an effort to find consistency between the broad, but crude, overview provided by seismic data with the much more detailed well data that come from a specific, pinpointed location of an oil well (see figure 5.4). There is rarely complete consistency between seismic and well data. Instead, consistency is *crafted* through labor-intensive work. As one of the explorationists tasked with a well tie-in and visually superimposing well data onto seismic data explained, "It does not fit." Still, he was not really despairing. Consistency is rarely the case because "it matters how old the wells are, what types of data were collected, how far away the wells are. If they are close, that is obviously beneficial." He continued working. The inconsistency between seismic and well data is compounded by the fact that they are measured with different scales. Seismic data are measured as the time that it takes an echo of a particular sound wave to travel back to the sensors after being refracted by subsurface rocks, while well data are measured relative to the depth measured

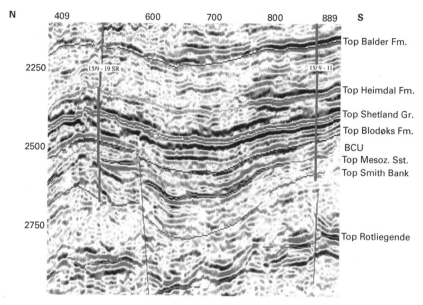

Figure 5.4

Input to well tie-in where two wells marked 15/9–19SR and 15/9–11 are calibrated against the seismic.

Source: Reproduced by permission from the Volve open data set.

in meters of the well where they were recorded. The heart of the problem stems from the fact that the speed of acoustic waves differs with different types of rocks, resulting in a nonlinear time-to-depth conversion. In addition, the explorationists work to gain a sense of subsurface *nonconformities*. Nonconformities result from geological processes such as fault lines. As our informant described, "If you have very steep non-conformities, [the nonconformities] can jump several hundred meters back and forth from time[-based measures] to depth[-based measures]."

DISCONTINUITY: CONTESTING THE ESTABLISHED

The set of explorationists' work practices illustrated above is, as the label of continuity signals, a conservative one in the sense that it is geared toward confirming a prediction—that is, supporting a particular prospect. By ironing out inconsistencies in the data, by elaborating and extending simulation models, by filling in gaps in the data by extrapolating from available data, by triangulating between independent data sets, and by engaging in peer-based deliberations, the explorationists gradually accumulate details of their prospect qua predication. In this manner the prospects in NorthOil's portfolio gain organizational credibility and move through the staged, formalized decision-making funnel model of figure 5.1. Albeit the dominant mode of working, this conservative mode of working is not the only one. It is at times punctuated by new data or modeling that can be neither accommodated nor dismissed, giving rise to the second mode of working for the explorationists.

With a bit of drama, one explorationist exclaimed, "Any new well can change the basin evolution; any new well can change our predictions!" He knew very well that he was exaggerating to make his point. A change in the basin evolution is a consequential change. The basin model that the explorationists rely on when searching in an area is effectively the prevailing understanding of that area's geological history. It is, as illustrated above, the result of extensive effort and represents a significant sunk investment in terms of earlier work. Part of the work practices above was substantiating a prospect through different types of simulation and modeling. A basin model is a model of the history of the area's geological evolution. Once a basin model is conceptualized, it is tested against existing well data for consistency. However, in practice, consistency is never fully achieved. Working on a basin model, one explorationist described how the team selects two hundred reference wells out of a sample of one thousand wells to support this consistency check. Well data are inconsistent, so they use heuristics such as the well's age (assuming new wells have better data quality than older ones) and then consider how much work went into calibrating the data, noting that often "we must go in and calibrate the well to the seismic [i.e., well tie-in]. And if it is a bad calibration, if things do not match, then the logs are poorly collected." Poor quality can be tied to a variety of reasons—for example, "things that happened on the rig that are not documented well enough, that give a sloppy [well] log."

The explorationists, out of necessity, harbor a high tolerance for inconsistencies. The sheer volume and types of imperfections in the available geodata preclude a complete cleansing despite the best efforts of data managers (see chapter 4). The corporate setting in which the explorationists operate is highly disciplined when it comes to reducing the efforts of sorting out ("academically interesting") inconsistencies and instead encourages living with them. If an explorationist tolerates many inconsistencies, this begs the question of what it will take to punctuate that tolerance; that is, under what conditions are the inconsistencies significant enough to make a difference? There is no formula here but rather a collective judgment reached by peer-based deliberation grounded in the cumulative weight of data and models at any given time.

Some inconsistencies are more dramatic because they suggest fundamental problems with existing models. They relate to situations in which one's concept, the play, is challenged. One example is when one explorationist was grappling with a particular form of inconsistency in her prospect. The seismic she was examining covered a large area but was coarse (see chapter 3). It showed sand throughout the field, but the well data told a different story: "I have a well here [pointing to her screen] that hits sand, and I have a well here [pointing] that does not hit sand. And then I have a seismic processing [pointing to another location on the screen] that shows me it should be sand all over. Then I need to decide: No, that [pointing] is not sand; this [pointing] is sand."

To account for different probabilities, data sometimes need to be extrapolated from geographic areas that are less known into geographic areas that are better known geologically (e.g., more wells have been drilled or more seismic surveys have been shot). Data are then extrapolated, as one explorationist explained, talking about a well: "Ok that one, it can be very far off, ok, the data of the well is put in here. If I were right and we are at the same time and in the same kind of rock etc. I take this well and I put it here and say, I use this porosity, I use this permeability. As an analogue."

A principal trigger for challenging or contesting existing prospects is the arrival of new data. Explorationists are eager to get their hands on this data. Well data, with its fine-grained measurements, are particularly appreciated. Since seismic data are coarse-grained, well data are the closest that explorationists come to hard evidence. Given the considerable financial costs of drilling new wells, NorthOil invests in the drilling of a few dozen in a typical year in the area reported on in our case study. The explorationists' appreciation for new well data leads them to cut corners in formal procedures. Rather than use the formal, time-consuming process to ensure quality-controlled data from an ongoing well-drilling operation lasting a month or two, they import the data directly from the drilling database. After years moving through the stages of the funnel model (figure 5.1), the drilling data will finally give feedback on the explorationists' predictions: Was there an oil discovery as they had predicted? However, new data also provide a much-appreciated occasion to consider unproven plays and alternative geological scenarios: "When we have a new well, it is not like we do not care anymore [whether there was oil or not]," one explorationist told us. "We use it for future exploration. . . . I care about the data. Data from the well is key."

The explorationists will at any point in time entertain—and generate—a number of leads, or ideas, that might later be turned into full-fledged prospects. Engaging leads is thus a way to challenge or contest existing prospects as well as generate new ones. At one level, coming up with new leads is a continuous activity: "We generate [leads] all the time, as many as we can. And it does not need to take any more than a few days. We can approve them ourselves as the parameters are loose and not particularly precise." However, the explorationists have internalized the operational reality of a highly competitive business environment. New data come with a hefty price tag. Well data from drilling, in particular, but also new seismic surveys represent significant economic investment. Searching for new leads and prospects is, accordingly, directional and goal-seeking rather than open-ended searching. Resources always constrain the search. They are never exhaustive or perfect. As one explorationist pointed out, "We often do not have time to work out all [the concepts]; it takes too much time. We very often have limited time to drive concepts forward. It can be a matter of a few months." The resource-demanding nature of assessing the credibility of new concepts forces good-enough, pieced-together rather than elaborate assessments, as the same explorationist went on to explain: "A lot of data must be pieced together, [and] a model needs to be built and to run basin simulations. In sum, it is a bit hard."

MULTIPLICITY: IRREDUCIBLE AND CULTIVATED

The abductive nature of pragmatic action captures well the time- and resource-bounded constraints of explorationists' procedures for making predictions (prospects, leads), as illustrated by the above modes of working. Compounded by the business environment they operate in, their predictions are "good enough" to satisfy the institutionalized decision-making processes of the funnel model (March 1994). The funnel model, as prospects get close to being considered drilled, need to be equipped with quantified estimates of the volume, risk, and value of a prospect. However, the version fed to management ("the guys upstairs") for formal decision-making radically under-communicates the prevalence of the multiple competing possibilities known to the explorationists. For purposes of arriving at a managerial decision, the extent and role of multiple divergent predictions are bracketed. However, among the explorationists there is an openness to entertain multiple incompatible possibilities at the same time.

As mentioned, the explorationists have a deep appreciation for new data, and not the least well data, the closest they come to hard data. This, however, should not be misconstrued to suggest that explorationists trust well data and take it at face value, as such data regularly provide deeply ambiguous results that feed divergent possibilities, none of which can be put to rest by the data themselves. One explorationist's struggle to make sense of his well data illustrates the irreducible—even in the presence of the hardest data they have, well data—interpretations they need to entertain. This explorationist was studying the analysis of the chemical composition of the hydrocarbons from a particular well. Such an analysis, while not quite as accurate as DNA profiling, is still useful and instructive because every oil reservoir has hydrocarbons with a distinct chemical profile that should allow one to differentiate hydrocarbons from two different oil reservoirs. Normally, you would assume that two wells near each other would draw from the same oil reserve. This is the why the explorationist was puzzled: "In one of the fields in our area, each well is different when it comes to the [origins of the] hydrocarbons. They have different chemical compositions, which is really strange. They are so close by, you would think they are all the same, but they are not. [The geology] is very complex in some areas." His reference to "complex" geology was to say that there are multiple irreconcilable—given the available data—interpretations of the geological narrative. This ambiguity or multiplicity is not so much resolved as relegated to a nagging uncertainty that, in later situations, may turn into a salient, rather than a latent, possibility.

Digital tools are invaluable for the explorationists. They have a wealth of different tools that help them manage and interpret well logs, process and interpret seismic data, and conduct seismic well tie-ins as well as basin modeling and simulation (see table 4.1). Increased computing power coupled with new methods for seismic processing makes 3D seismic cubes (i.e., 3D seismic data sets synthesized from multiple sources; see chapter 3), previously prohibitively time-consuming, more practical. As a consequence, many more than those actively pursued tend to be generated, promoting questions about which one to lean on (see chapter 4). For instance, one explorationist described over lunch one day how he was working in a field with two hundred variants of the same 3D seismic cube. There was no way of discerning the purpose of all two hundred variants. The one officially quality-controlled variant shed little light on the others. As he was interested in a particular subsurface level in the project, he investigated it. Perhaps there was an underlying, implicitly assumed idea that he had missed, he asked himself: "What was this idea? Why? It is not apparent in that 'pick' [their term, implying interpretation of a subsurface of the seismic level in light of subsurfaces picked from well data]. You have some new data that do not fit. How does it relate?"

Coping with multiple possibilities is fundamentally collective. In formal but, importantly, more often informal peer-based discussions, explorationists collectively deliberate multiple possibilities: "When you talk to experts and advisors, they stress the nuances, and the details in it, and not at least the dimensions in it." Peer-based discussions are vital to avoid the tunnel vision that working strenuously with a prospect might easily create. The prevalence of multiple, as well as radically different, interpretations is internalized by explorationists as part of their professional identity. However, institutional constraints make it organizationally and politically necessary at times to bracket this inherent multiplicity. Multiplicity is not resolved or eliminated as much as put temporarily on hold for the purposes of passing one of the funnel model's decision gates. The task of "risking" (their term) a prospect is illustrative. Risking is the quantification of qualitatively manipulating the prospect. One explorationist commented on the assumptions underpinning risking: "We put a [quantified] probability that you have a trap, that it is sealed that you have a reservoir, that you have migration." Crucially, you assign quantified measures for variables such as rock porosity and permeability, oil saturation, viscosity, and volumes to your prospect. Despite estimates, risking contains "a lot of speculation and [subjective] opinions" but still necessarily legitimizes NorthOil's gated decision processes. The problems with quantifying the probabilities of an oil discovery for different prospects under consideration are particularly pronounced for those with medium-range probabilities—that is, 10–25 percent: "Here we are struggling. They diverge in all directions."

CONCLUSION

The geological structures that the explorationists work with vividly illustrate the liquefaction, or disembedding, of digital representations from their originating physical referent discussed in chapter 1. For all practical purposes, the everyday work of explorationists is with IoT-rendered representations of the subsurface, not the physical geological structures kilometers below the seabed. This is the sense in which this book responds to Boelstorff's (2016) call to focus on how the digital can be "real" rather than maintain a dichotomous separation of the physical/real versus the digital/virtual: digital representations are real in their implications for work practices and knowing. As the present chapter makes clear, in everyday work, digital representations "stand in for"—are—the geological subsurface (Bailey et al. 2012; Leonardi 2012).

The objects of explorationists' knowing correspond to Knorr Cetina's (2001, 190) notion of epistemic objects that are never stable or fixed but rather "are more like open drawers filled with folders extending indefinitely into the depth of a dark closet. . . . They continually acquire new properties and change the ones they have" (cf. Kallinikos et al. 2013). The three modes of explorationists' grappling with the uncertainties of their epistemic objects, analytically separated but empirically summarized in table 5.1, address the fundamental tension underscored by Rheinberger (1997, 80): "To remain productive in an epistemic sense, an experimental system must be sufficiently open to generate unprecedented events. . . . At the same time it must be sufficiently closed to prevent a breakdown of its reproductive coherence."

Table 5.1
Overview of the three analytic modes of explorationists' work practices when grappling with geodata.

Mode	Characteristics
Continuity	- Ironing out or dismissing wrinkles - Gradual refinement by triangulation - Looking for gaps in the data and filling these
Discontinuity	- Confronting and resolving inconsistent, especially new, data - Engage with possible leads - Do just enough model revision to accommodate data
Multiplicity	- Embrace multiple divergent models - Collective, peer-based sensemaking

The first mode of explorationists' working (continuity) is about confirming and supporting an existing prospect. The explorationists accumulate support or evidence for their prospect by resolving inconsistencies, filling in gaps, and triangulating across data sets. Crucially, there are different levels of uncertainty (Chang 2004; Shapin 2011; Østerlie and Monteiro 2020). For instance, well data measured along a drilled well in precisely one location is viewed as more reliable than the coarse seismic data, even though the seismic data cover several square kilometers (see chapter 3). The example above with well tie-ins illustrates how the "hard(er)" evidence of well data is used to calibrate the seismic data.

The modus operandi of the first mode of continuity is that of conservatively confirming an existing prediction. The cumulatively increased sunk investments in terms of effort, resources, and time risk creating path dependencies as infrastructure studies demonstrate. The second mode of explorationists' work practices (discontinuity) is a counterreaction. This abductively challenges and contests the former mode. Efforts to iron out wrinkles, inconsistencies, and outliers are attainable only to a certain level. The arrival of new data sets (e.g., the drilling of a new well) may trigger abductively searching for new ways to make sense of the data, new and old. It amounts to coming up with a new geological narrative (see Wylie 2017) that reframes earlier prospects.

The third mode of working (multiplicity) is different. It addresses how explorationists cultivate and encourage—embrace, rather than eliminate—epistemic uncertainty while remaining sensitive to the corporate necessity of not halting operations. The well-defined search for proven plays may, and regularly does, spill over into the ill-defined search for unproven ones (i.e., potential, but not yet demonstrated, geological configurations for geological traps). This mode of working fills productive, organizational roles, as it is "through divergent or misaligned understandings that problematic situations can give way to positive reconstructions" (Stark 2011, 192). The multiplicity of interpretations is regulated collectively. They are deliberated in collective arenas that yield partial agreements (Oborn et al. 2011). Consensus is thus never arrived at but rather worked out through temporal and local arrangements, resonating with Mol's (2003) study of how medical specializations such as surgery and pathology—despite radical differences in routines, theories, vocabulary, and instruments—forge temporary agreements about how to treat atherosclerosis in patients.

6 KNOWING

written with Thomas Østerlie

One must therefore know the *method of knowing* in order to grasp the *object to be known*.

—Bachelard (1949/1998, cited in Rheinberger 2005)

The empirical focus of this chapter is on sand. With many of the hydrocarbons on the Norwegian continental shelf trapped in Jurassic sandstone, the risk of sand in the production flow is immanent. Entering the production system through wells drilled thousands of meters into the earth's crust, the fluid rushing into the well sweeps sand with it along the pipelines all the way to the topside processing plant, where sand settles in the tanks that separate crude oil and natural gas from the other constituents of the fluids. Sand deposits threaten to reduce the plant's processing capacity and oil quality. More importantly, however, sand particles rushing at high speeds through the pipelines erode the piping, eating away at the valves controlling the fluid flow as well as the valve casings. Left unchecked, high-speed fluids, gases, and sand particles jetting out of a puncture may cause catastrophic environmental damage as oil gushes into the ocean, while leaking gas carries a danger of igniting and exploding on the topside platform.[1] It is therefore important for the offshore control room operators to take mitigating measures to prevent sand from entering the production system.

Sand-monitoring rounds are traditionally part of the offshore roughnecks' daily inspections at the offshore processing plant. When roughnecks discover sand in the production equipment, offshore laboratory assistants embark on a regime of inspecting and emptying the sand traps—cups mounted underneath each flow line where the heavier sand particles settle as fluids rush past—to locate the originating well. It takes time for sand deposits to accumulate in the sand traps, though, so laboratory assistants can only inspect the cups once during every eight-hour shift. It may take days or even weeks before the well is back in production without new sand entering it.

Sand-monitoring routines were targeted or digitalization for reasons of efficiency, quality, and safety. The manual routines are labor-intensive, error prone, and time-consuming. Responding to business pressure, oil operators on the Norwegian continental shelf, including but not limited to NorthOil, are continuously engaged in cost cutting. An important, and for sand-monitoring routines relevant, result of this is to shift work (and workers) from offshore to onshore. Offshore workers on the Norwegian continental shelf have negotiated a two-weeks-on, four-weeks-off work schedule in addition to offshore salary bonuses. With most if not all offshore workers globally working two weeks on, two off, oil operators are constantly looking for ways to shift tasks onshore or automate them altogether through digitalization. As pointed out in chapter 1, removing manual tasks and workers from offshore installations has a long, ongoing history despite warnings about eroding safety from labor unions and the Petroleum Safety Authority Norway (see also Ryggvik 2018).

In the context of this book, the case of digitalizing sand-monitoring routines is illuminating. It ties directly into the fundamental discussion raised in chapter 1 about the conditions under which the representational capacity of digital data holds organizational consequences (Kallinikos 2007). Through a series of efforts, NorthOil explored how different digital representations—Internet of Things (IoT)–based sensor measurements, graphs, plots of historic data, and predictive simulation models—attempted to stand in for the all-too-physical reality of sand eroding the pipes, chokes, and valves of the oil production facility (Leonardi 2012).

A sequence of digital renderings of sand was successively superimposed onto physical sand, traditionally collected in cups and analyzed in laboratories. What, then, in the everyday practices of sand monitoring is "sand"? Closer to the heart of this book, what role do the different digital renderings of sand play in the transformation of knowledge-based practices of sand monitoring? In other words, I am focused more on the (epistemological) concern of how sand monitoring is achieved than the (ontological) concerns about what sand really is. When digitally transforming work practices rely on a multitude of connotations of sand, be it physical detected sand deposits in sand traps, sensor readings, or historic and projected plots, how do operators know and act upon sand? How do operators know—enough—about sand to take mitigating actions, such as ordering maintenance interventions, implementing supplementary inspections, or, ultimately, reducing production capacity?

Anything but reified, knowing underpins action/practice (Alavi and Leidner 2001). Knowing, Orlikowski (2002) notes, is "a situated knowing constituted by a person acting in a particular setting and engaging aspects of the self, the body, and the physical and social worlds" (252). However, as influential insights in the social sciences have made clear during the last couple of decades, all knowing practices are material (Barad 2003; Cecez-Kecmanovic et al. 2014; Latour 1999; Orlikowski and Scott 2008). There is thus a broad consensus *that* knowing is material but a significant divergence regarding *how*, be it "entangled" (Orlikowski 2006), "imbricated" (Leonardi 2013), or "inscribed" (Monteiro and Hanseth 1996). The challenge is to specify, in interesting detail, how the knowing of digital sand is done—that is, unpack the underpinning empirical conditions and mechanisms.

The versatility of digital technologies relies on the capacity to digitally represent and subsequently algorithmically manipulate selected physical processes, objects, or qualities within a domain (see, e.g., the key role of sensors and IoT pointed out in chapter 1). How closely the digital representations mimic the physical domain varies from directly mirroring, to resembling, to decoupled. Pressing the capacity of digital representations to decouple as much as possible is important because this "has the greatest potential to change work's historically tight coupling to the physical and, with it, the work relations of people to objects and each other" (Bailey et al. 2012, 1486). In other words, the disruptive potential of digital technologies assumes the capacity of digital representations to decouple from, not merely mirror, existing work practices (Borgmann 1999).

Several decades of empirical studies of digital technologies in organizations, however, demonstrate how technological potential often fails to translate into organizational change in practice (Zuboff 1988; Leonardi 2012). For digital representations to underpin organizational change in practice, they need to be implicated in consequential decisions and actions within work practices. To become *organizationally real*, digital representations, beyond their mere potential/capacity for decoupling, need to be incorporated into organizational practices; digital representations are not, but may become, organizationally real. This entails that the focus is on the conditions and mechanisms through which digital representations get woven into institutionalized work practices of sand monitoring.

Thus, the empirical case of successive stages of digital sand presented below is not one of disruptive change. On the contrary, it is an account of the gradual institutionalization of digital representations, increasingly distant from "real" sand, into work practices. A crucial aspect of the case is the manner in which a singular digital representation, in and of itself, carries little weight; to carry more weight, it needs to be tied into a broader set of supporting digital representations, a *machinery*.

THEORETICAL FRAMING: AUTOMATION AND IOT-ENABLED VISIONS OF INDUSTRY 4.0

Visions and proclamations of the Second Machine Age, or Industry 4.0, or the Industrial IoT draw heavily on rich yet underspecified accounts of what digitalization is and entails. Enabling technologies such as cloud computing, big data/analytics, robotics, and the IoT are regularly identified but without adequate explanation as to where, when, and how digitalization unfolds.

Historically, digitalization has been tied to automation, the substitution of previously manual tasks for digitalized ones.[2] The introduction of computers in the workplace in the 1970s, 1980s, and part of the 1990s regularly spawned fears of job loss and deskilling among employees (Braverman 1974; Friedman 1977). In Europe, more than the US, unions mobilized to respond to these perceived threats. In some countries, this resulted in new legislation and regulations ensuring employees' right to consultation or participation when computers are introduced in organizations (Asaro 2000; Muller and Kuhn 1993).

Conceptualized as computerization, the introduction of digital technologies was historically tied to their potential to automate a wide set of work tasks (Friedman and Cornford 1989). Braverman (1974) argued in an influential study that the scope of computerization would imply widespread deskilling of work tasks. The defining assumption, automation by substituting for manual work tasks, met with growing critique of both an empirical and theoretical nature.

Empirically, scholars demonstrated that the results of computerization were significantly more varied than what Braverman maintained. Barley (1986), for instance, showed how the introduction of similar computed tomography scanners in different hospitals led to different work routines and roles for radiologists. Similarly, the coining of the so-called productivity paradox underscored the variations in outcomes of computerization: studies found negative, zero, and positive correlation between investments in computers and productivity (Kling 1996). A series of studies on computerization demonstrated that digital technologies involved local appropriation and hence were not merely automation (see, for instance, DeSanctis and Poole 1994).

Theoretically, the variance in empirical results of computerization led to identifying an assumption of technological determinism in Braverman. It was, accordingly, necessary to establish the significance of digital technologies as something different from (only) automation i.e., that the dynamics around the development, use and subsequent spread of digital technologies differ from those predominately addressing automation. So how and why, then, are digital technologies different?

In Zuboff's (1988) formulation, digital technologies' potential for transformation was unique as, beyond automation, they had the ability to informate (see discussion in chapter 1). Informating relies on a "spillover" effect in digital technologies—that is, data input to processes and tasks is not consumed. The fundamental insight of Zuboff's notion of informating was to underscore the inherently open-ended, unfinished, and extendable character of digital technologies. This has been incorporated into more recent conceptualizations of digitalization (see Garud et al. 2008; Kallinikos et al. 2013) as digital technologies are inherently dynamic and malleable (Yoo et al. 2010).

Scholars of digitalization, albeit from different angles and with different formulations, provide strikingly similar insights: Yoo et al. (2010) identify the defining quality of digital technologies, their (algorithmic) programmability and layering; Zittrain (2006) characterizes the open-ended extendibility of digital technology via the notion of "generative"; Lusch and Nambisan (2015, 160) identify the defining ability of "liquefaction" of digital representation decoupled "from its physical device or form" (see chapter 1); and Borgmann (1999) notes the ability of digital representations to "illuminate, transform, or displace reality . . . [and hence] disclose what is distant in space and remote in time" (1).

Thus, to talk of tools and technologies mediating the outside world downplays to a level of nonexistence the active contribution of the tool/technology. Breaking away from a representational perspective where reality is passively mediated by tools and technologies (Pickering 2010; Jones 2019), a performative perspective underscores their coconstitutive relationship (MacKenzie 2006). In a widely cited study of the financial option market, MacKenzie and Millo (2003) explicitly set out to demonstrate the performativity of the so-called Black-Scholes model by showing how its initially descriptive role gradually got replaced by an enacting role when the formula was inscribed in trading robots and professional routines. As MacKenzie and Millo (2003, 107) note: "Option pricing theory . . . succeeded empirically not because it discovered pre-existing price patterns but because markets changed in ways that made its assumptions more accurate and because the theory was used in arbitrage" (cited in Orlikowski and Scott 2008, 461). The crucial relevance for the argument in this book is that knowing sand is inherently caught up in the sociomaterial means of knowing sand; *what* operators know about sand is *how* they know it (Rheinberger 2005).

The above outlined theoretical interest in characterizing digitalization is radically boosted by the empirical emergence of big data together with data-driven, machine learning–based forms of algorithmic manipulation. Socially informed critical studies of digitalization, to further our understanding, need to combine a theoretical grasp of digitalization with an empirical grounding in organizational dynamics, a combination largely missing when it comes to the data-driven algorithmic approaches literature reviews consistently find (Günther et al. 2017; Sivarajah et al. 2017). A particularly helpful approach is provided by Bailey et al. (2012). They analyze the *degree* to which digital representations are disembedded from their referent. This paves the way for inquiries, like the present one, about the conditions under which (degrees of) disembedded digital representations carry weight. Drawing on the semiotics of Peirce, Bailey et al. identify three configurations of the digital representation/physical referent relationship: (1) indices, where digital representations are but labels for their physical referents, the desktop metaphor of graphic interfaces being an example; (2) icons, where the digital representations are similar but not the same—for example, a videoconference instead of a face-to-face meeting; (3) symbols, where the digital representations bear no resemblance to the physical referent, and the link is solely based on conventions.[3]

Taking the digitalization qua liquefaction to its limits, some studies focus on simulation as explorative—a radical break from existing work routines. For instance, Dodgson et al. (2013) analyze how simulations are used in a business organization to promote "processes that induce and sustain the craziness of wild ideas" (1359; see also Dodgson et al. (2007), exploring radical design changes). As pointed out above, pursuing the potential of digitalization, in principle, as completely disembedded from the referent sidesteps the crucial concern of how, in practice, to give simulations organizational weight. The latter inevitably involves focusing on the relationship of digital representations to their referents.

One stream of work focuses on the dangers of simulations replacing their referents. Turkle's (2009) work emphasizes the dangers of simulation-based renditions of reality given their strong, seductive capabilities. As users are gradually immersed in simulations, "Familiarity with the behaviour of [digital representations] can grow into something akin to trusting them, a new kind of witnessing" (Turkle 2009, 63). The physical referents are central to Turkle's analysis of simulations to the extent that she warns of the dangers of their disappearance.

Leonardi's (2012) work on the automotive industry's attempt to replace physical (and costly) car crashes in safety design models with simulated crashes is one of the few longitudinal studies of simulations in organizations. A principal finding is that simulations, to carry organizational weight, need to be icons rather than symbols. As Bailey et al. (2012), drawing on Leonardi's study, point out, "This tight coupling in simulation means that people who create representations are highly dependent on physical referents" (1500). Phrased in Peirce's vocabulary, Leonardi (2012) argues how simulations can avoid being empty representations (symbols) and gain organizational relevance by becoming icons—that is, enjoy immediate recognition through their "similarity with the [physical] object" as "seeing is believing" (14–16).

Consistent with insights from infrastructure studies, however, singular digital representations of sand carry little weight. To carry weight, knowledge about sand needs an accompanying infrastructure of vocabularies, institutions, practices, and technologies (Poovey 1998; Porter 1996). With science and technology studies (STS) underscoring the local, embodied, and enacted nature of knowing "facts," the ability of knowing sand to "travel" from the offshore installation to the onshore operations center may seem a paradox: "If, as empirical research securely establishes, science is a local product, how does it travel with what seems to be unique efficiency?" (Shapin 1995, 307). In the colorful phrasing of Morgan (2010), the apparent paradox is resolved as knowing sand and "facts require[s] a variety of charismatic companions and good authorities to travel well" as "we depend upon systems, conventions, authorities and all sorts of good companions to get facts to travel well—in various senses—and danger may lurk when these are subverted or fail to work" because "once facts leave home, it is more difficult to keep them safe" (4–6). As will be demonstrated empirically below, facts and knowledge about sand, expressed in a variety of digital representations of sand, also need traveling companions to carry organizational weight in NorthOil.

THE CASE OF SAND-MONITORING ROUTINES:
THE SUPERIMPOSING OF DIGITAL REPRESENTATIONS

The setting for the case below on successive stages of digitalizing sand monitoring—digital sand—is the onshore operation center for daily production and its interactions with the offshore control room at the field in question on the Norwegian continental shelf. In parallel, corporate research and development (R&D) engineers in close collaboration with technology vendors, producers of sensors, and a standardization organization are engaged in a digitalization effort to improve (in terms of quality, efficiency, and safety) sand monitoring in which they, in addition to field testing, organize workshops and meetings to discuss and explore alternative approaches to digital sand.

As such, the case was an early attempt at what is now a proliferation of remotely operated onshore control centers being delegated increasingly more tasks and responsibilities (see Latour [1987]'s notion of centers of calculation). Remotely operated control centers come in different and evolving configurations (Hepsø and Monteiro 2021). Some, like the sand-monitoring case below, set up corporate control rooms that are assigned a particular field with associated "satellites" (additional, neighboring fields tied into the same processing and transportation infrastructure as the original ones, thus benefiting from the sunk investment). Others act as expert or consultation centers serving a group of oil fields. Yet others are run from the premises of the service provider, not the oil operator. In sum, this has resulted in the currently established dominant operational model of offshore installations being minimally manned or unmanned. Configurations of remotely operated control centers are thus vital ways in which offshore oil on the Norwegian continental shelf is pushing toward visions of automated oil production.

In what follows, a longitudinal account of the digital transformation of sand monitoring over a period of almost twenty years, from the mid-1990s to 2012, is presented. It sheds light on how, if at all, digital representations become organizationally real. Through a succession of different initiatives in response to setbacks and challenges, digital representations of sand are successively superimposed onto physical sand; gradually, digital sand becomes organizationally real.

The transformation of sand-monitoring practices is traced through four periods during which configurations of material arrangements (typically sensors but also the physical equipment associated with leading fluids from the wells) and digital algorithms (for transforming sensor output into different digital data) successively disembed physical sand (the referent) into digital representations (the reference; see table 6.1). With the successive

Table 6.1

An overview of the different periods of digital sand monitoring.

Period	Focus	Key technologies	Central actors
Period 1 (early 1990s): organizational consequences of false alarms	Testing of sand-monitoring measuring devices (sensors). Struggling with proliferation of false alarms that underpin confidence hence relevance of the sensors.	Singular, stand-alone sensors (electrical resistance). No algorithmic transformation beyond monitoring threshold values.	Offshore control room operators and onshore production engineers. Dialogue with vendor of sensors to increase accuracy.
Period 2 (mid-1990s–early 2000s): meshing sand monitoring with related routines	Real-time measurement of sand content as characteristic of the well flow to replace manual and time-consuming sand-monitoring practices. Operational principle of zero sand tolerance implemented as immediate shut-in of sanding well.	Digital sand sensors (acoustic, electrical resistance). Algorithms transforming acoustic/electrical resistance data into measure of sand content in well flow. Simple user interface to sensor measurements.	Digital sand mitigation within offshore control room operators' production control practices for minimizing disruptions to offshore processing plant. Onshore production engineers supported control room operators in investigating sand alarms.
Period 3 (early–mid-2000s): sand, an interactive, algorithmic phenomenon	Combining real-time measurements with geomechanical theory and with geomechanical knowledge on causes of sand influx transforms digital sand from a characteristic of well flow to an indicator of events in the reservoir. Zero sand tolerance policy implemented by sand mitigation strategies fitted with the kinds of sand events causing sand influx.	Visualizations of sand content data development over time in trends. Inclusion of data points from other sensors (temperature, pressure) to better identify false alarms. Dashboard aggregating alarms across all wells on the oil field.	Sand mitigation nominally within offshore control room operators' production control practices. Sand mitigation handled in practice by onshore production engineers to limit the impact of sand incidents on optimizing daily production volumes.
Period 4 (mid-2000s and onward): sand as a machine learning–based predictive model	From zero sand tolerance policy toward predicting the effect of producing with limited amounts of sand in the well flow.	Algorithm for predicting erosion on pipeline bends and valves.	Sand monitoring exclusively in the domain of optimizing daily production volumes by coordinating erosion of production equipment with maintenance shutdowns of the offshore plant.

disembedding of physical sand throughout these four periods, the refer-ent/reference relation expands the scope of digital representations from a real-time measurement of sand influx in individual wells to simulations of sand erosion in the production equipment. The character and focus of sand-monitoring practices transform with the different renderings of digital sand.

PERIOD 1 (EARLY 1990S): ORGANIZATIONAL CONSEQUENCES OF FALSE ALARMS

"So, you may have sand in the production system," explains NorthOil's leading expert on sand mitigation technologies. He is giving a presentation at one of the many workshops of the joint industrial R&D project. The undersized meeting room is stiflingly hot, with between ten and fifteen people in attendance cramped around a too-small conference table. Still, representatives from most of the major operators and leading digital sensor technology vendors in the North Sea region are present. During this pre-sentation, the sand expert—fondly referred to as Mr. Sandman within the national petroleum industry—reflects on some of the corporation's earliest experiences with introducing a digital sand-monitoring system for use by offshore control room operators in the mid-1990s.

"And these are the [sand] data values," he continues as he draws a jagged, rising line with a black marker in a coordinate system on the whiteboard. Still with his back turned to the rest of us attending his presentation, he picks up the red marker. "And then there is this one single peak above the alarm limit," he says as he draws a horizontal red line across the coordinate system to indicate the alarm level, "and then you have triggered an alarm in the offshore control room's process control system. That's just stupid!" He turns toward the rest of us, pausing to emphasize his point.

Some workshop participants nod in agreement. Others murmur, pos-sibly in consent, or maybe just to indicate they are paying attention.

After a pregnant pause, the sand expert continues: "What happened, you see, was that they [the offshore control room operators] ignored the alarms, and they said"—and now he speaks with a theatrically exasperated voice, paraphrasing the control room operators—"'The system you have is rubbish [and] we are not able to monitor for sand influx.' So, they turned it off, never to use it again."

What the sand mitigation expert is vividly demonstrating is the organizational consequences of false alarms. The offshore control room is a hectic place. The two operators working there have to be vigilant at all times to make quick decisions in response to audio alarms going off when key operating parameters exceed preset alarm limits—and alarms go off on a regular basis. If you have sand sensors with "one single peak" triggering a full control room alarm, it is disruptive to daily control room operations. Sensors, sand sensors included, are notoriously error prone and therefore regularly produce outliers that the operators have to ignore. To be of practical use, sand-monitoring systems need to be less brittle, less sensitive to noise, and hence more robust. Failing in this regard, the result was that the complete sand-monitoring system was turned off. These early experiences with sand sensors provide important empirical elaborations on the capacity of liquefaction discussed in chapter 1—that is, the disembedding of the sand sensor measurements from the physical presence of sand in the production flow. Despite a capacity, it is not realized organizationally; it never becomes organizationally real.

The petrochemical-processing plant is a controlled system, very much akin to a laboratory, being designed as a closed-loop system. The term "closed-loop system" stems from cybernetic theory, which describes it as a controlled system that deterministically transforms input to output. As a closed-loop system, the petrochemical-processing plant is physically constructed as a series of subsystems with fluids streaming through and with each subsystem in itself a closed-loop system. Physical barriers prevent possible instabilities and fluctuations within one subsystem from spreading throughout the entire processing plant. Dampening fluctuations this way, the physical setup creates a deterministic system. When an operating variable such as pressure increases or decreases at a point in the petrochemical-processing plant, it does so linearly. All the control room operator needs to see is that the figure increases or decreases. In this laboratory-like environment, there are no intervening variables interfering with measurements.

The well flow, in contrast, is by no means as well behaved. For the sand sensor in question, the vendor had tested and verified the digital sand-monitoring system under laboratory conditions. Hence, the sand data help in the sense of yielding adequately accurate laboratory results. The vendor in dialogue with the technical division of NorthOil agreed that these tests fulfilled the requirements set by NorthOil. When mounted within pipelines embedded deep within the earth's crust, however, multiple factors registered as increasing sand content. Furthermore, unlike the processing plant the rocks and fluids within the reservoir were by no means under cybernetic control. Developments within the reservoir therefore did not register as linear or deterministic changes in the sand data. Despite being subjected to rigorous testing and technical qualifications, the credibility of digital sand for this sand sensor failed to hold up in production settings beyond laboratory conditions.

The offshore control room operators could not live with the resulting level of false alarms. Their primary concern is to ensure stable, uninterrupted operations of the topside processing plant. The rationale for introducing the digital sand-monitoring system was to reduce production loss due to sand mitigation. For the control room operators, though, reduced production volumes were of limited concern; their concern was and remains stable operations of the offshore production plant. Compared with the deposits of reeking, viscous, tarry sand accumulating in the topside production equipment, sand alarms turned out to be mostly false. In the time-pressured setting of the control room, the operators simply had no time to investigate whether triggered sand alarms were false or not.

PERIOD 2 (MID-1990S TO EARLY 2000S): MESHING SAND MONITORING WITH RELATED ROUTINES

Despite this setback, NorthOil pushed forward with its effort to introduce digital sand monitoring in the production organization. A few years later, the digital sand-monitoring system was in full use. By that time, however, the system had undergone several revisions. In addition, circumstances had changed in important ways. With a growing number of aging oil fields in the North Sea during the early 1990s, NorthOil—along with other operators in the region—was experiencing an increasing frequency in production disruptions because of sand in the production system. Ineffective manual routines for detecting sand entering the production system were identified as a principal contributor to significantly reduced production volumes. Thus, the prospect of an IoT-based, real-time rendering of digital sand, making it possible to identify and intervene immediately to stop the influx of sand, appeared increasingly attractive from a business perspective.

With renewed urgency, NorthOil mobilized many of the specialist communities previously involved in developing the original version of the sand-monitoring system to discuss how the data produced by digital sand sensors could be institutionalized into everyday practices, not become yet another technological pilot that gets abandoned soon after leaving the sheltered environment of the laboratory. As a result, digital sand monitoring was woven into the larger sociomaterial network of the offshore/onshore production organization to generate new data streams with associated arrangements for producing and handling these data streams, actualizing onshore production engineers' existing knowledge work in new ways to make real-time sand data actionable. This work spanned many settings and different regimes for making digital representations of sand robust.

In developing the digital sand-monitoring system implemented in the offshore control room, first these specialist communities focused on ensuring a "faithful" (see chapter 1) robust one-to-one reference/referent relationship between sand and its digital representation. Several research communities and vendor companies explored various possible technologies for doing this. With IoT-based sensing expanding in domains other than sand monitoring, they also considered repurposing existing sensors. The lead software engineer with the major vendor of digital sand-monitoring systems recalled: "We already had technology for inspecting pipeline integrity. When we saw the tender for a digital sand-monitoring system, we asked ourselves if our existing technology could also be used to detect sand in the well flow."

Measuring sand content required not only the development of digital sensor technologies but also the development of a standardized measuring scale. An international standardization organization had been tasked with developing a representative and reliable scale for the measurement of sand content, a standard that was missing. Standardizing sand content measurements, however, proved cumbersome. The problem, as O'Connell (1993) points out, was to find a *method* of measuring that was not (too) sensitive to the circumstances of its application. In the case of sand sensors, two different types of probes (i.e., sensors) were designed and tested, each tied to a different method of measuring sand. One was acoustic. It was based on picking up the acoustics of grains of sand hitting a surface and emitting a sound. The other was based on principles of electrical resistance. It was based on Ohm's law and measured changes in electrical resistance across a series of metal probes. Sand erodes the metal probes, causing a change in electrical resistance. This is transformed into a measure of sand content.

The problem facing both the proposed sand sensors, however, was their sensitivity to other phenomena or aspects of the environment than the intended one, sand. For instance, as a result of being designed around a hypersensitive microphone, the acoustic sand sensor registered all sound changes in the well flow, including the din of the production machinery traveling throughout the pipeline system. Likewise, changes in well flow temperature induce changes similar to those of erosion in resistance and will register as sand even if no sand is in the well flow. A senior engineer with the standardization organization summarized the situation during a project workshop:

> The specific way of measuring sand depends on a number of factors. For instance, different approaches are influenced by different factors such as pressure. We tried several approaches, but in the end, we arrived at the simplest way of measuring sand content: that of grains of sand flowing across a sensing probe every second.

The focus of these efforts was, again, on developing a credible measurement understood in terms of a faithful digital rendering of sand. Through testing and experimentation in their laboratory setting, the standardization organization's research engineers struggled to maintain the faithfulness of the measurement under differing conditions. In the end, settling on measuring sand content as the number of grains flowing across a point in space per second is the product of the research engineers prodding and tweaking the material arrangements to find the most robust relationship between sandy fluids streaming into a well and sand influx as measurable characteristics of the well flow.

During this, not only *what* constituted sand (a sensor measurement, not the tactile manifestation of accumulated deposits of a reeking, black, viscous tarry mass in the processing equipment and sand traps) and *how* it was detected (measuring the number of grains of sand flowing per second, not taking samples with subsequent laboratory testing) but also *who* was doing sand monitoring changed. Instead of offshore roughnecks, the task was shifted to onshore operations centers.

With the delocalization of digital sand, onshore production engineers and their professional expertise were mobilized to help monitor and mitigate sand. Production engineers' pragmatic concern is to optimize the field's daily production volumes. Their daily tasks evolve around planning and prioritizing production to optimally utilize the offshore production plants' processing capacities. They, much like the explorationists described in chapters 4 and 5, draw upon intimate knowledge of individual wells: the particulars of their designs, their production history, and their idiosyncrasies. "If you only learn one thing from your stay here," a production engineer explained during the lunch break one day, "[it is that] a well is never simply a well." What he meant, he elaborated, is that a well is "only a word." All wells are "different beasts," and "it's our job to know all of them."

This specialized and intimate knowledge of the material basis of daily production became increasingly central when investigating the fluid relation between digital symbols and their references from the mid-1990s onward. Digital sand management was integrated with production engineers' daily work tasks within NorthOil's onshore production organization. Working primarily with planning activities, the production engineers had more time to investigate digital data and their credibility. Digital sand monitoring latched on to these developments as production engineers would draw upon their existing knowledge of wells and sensors to triangulate sensor data with other information to determine whether a sand alarm was really caused by sand in the well flow or some other factor. "I'm not entirely convinced this is sand," one of the production engineers commented after the control room had informed him of a sand alarm. "The [sand] probe in E-37 has been acting up ever since the well was offline for maintenance." Hence, the necessary organizational credibility of sand sensors was *crafted* through the production engineers' efforts to control the data quality and calibrate and triangulate sand sensor measurements. To support this new form of sand-monitoring practice with increased roles and significance for the production engineers, the vendor of the sensor changed the entire sensor design by adding an additional probe shielded from sand to generate a data stream immune to the effects of sand but sensitive to other factors that could register as sand. Such interferences may include changes in the well flow itself, such as temperature and flow rate, or may be generated by the sensor itself, either because it has been mounted incorrectly, is broken, or is starting to wear out. The sand-monitoring system's vendor prototyped a new sand-monitoring application, correlating sand measurements with influencing factors so the production engineers could sort out data interferences. Upon noting a sand alarm, the production engineers use this application to weed out intervening factors registering as sand.

The production engineers are also part of the extended network of professions and activities working tightly to maintain and operate the production facilities. The production engineers draw upon the output of many of these activities, along with their own intimate knowledge of individual wells and how they may affect adjacent wells, to prod and investigate the relationship between digital sand data and sand entering the wells. Another sand sensor vendor then used this information to develop an elaborate calibration procedure that the production organization integrated with existing testing procedures. This is what NorthOil's leading expert on sand mitigation technologies was discussing in his presentation of the company's earliest experiences with digital sand monitoring, described above, during period 1. Having gained the attendants' attention with his statement about the control room operators turning off the digital sand-monitoring system, "never to use it again," he explained: "As part of velocity testing of the well, we would inject a pre-determined amount of sand grains into the well flow and then measure the signal at different production rates to calibrate the sand sensor." Thus, by linking sand monitoring with other routines, in this case well testing, the sand sensor received their much-needed calibration to enhance their organizational credibility. Another part of the maintenance organization developed new routines for ensuring that the sensors are mounted correctly, and the vendor made minor modifications to the sensor design to make them harder to mount incorrectly. At the same time, the production engineers started keeping tabs on the state of individual sand sensors in a collection of different documents and spreadsheets. Consequently, the nature of the work of monitoring and mitigating sand in the production system, as well as the organizing of this work, transformed with the digitization of sand.

While the production engineers' sand-monitoring work came to be increasingly centered around digital representations, a notable feature of this transformation of sand monitoring was an accumulation of representations, both digital and physical. These representations are always all in play, as production engineers also draw upon the topside inspection procedures to verify whether there is sand in the well flow. When monitoring for sand, there is never a dichotomous separation of the physical and digital. Rather, the production engineers move seamlessly between the two, as illustrated by the following episode.

Matt, that day's on-call production engineer, received a phone call from the offshore control room. "We have sand deposits in the separator [part of the offshore production system]," the control room operator said. Matt seemed puzzled. Looking at the dashboard showing the status of recent sand alarms across the field, he said, "There have been no sand alarms." "But we *have* sand in the separator," the control room operator insisted. Matt cycled through screens in the sand-monitoring application, looking for possible indications of sand but finding none. Leaning across the table toward a set of sand samples the production engineers kept handy, Matt picked one up. The vial's label said "Silt." Holding it by its neck, Matt shook the vial, looking at the quality of the sand swirling within. "What kind of sand is it that you've found?" he asked. "Silt," stated the control room operator. "Ah," Matt said, sounding relieved: "Silt is too fine to register on our sand sensors. There's no erosion danger, but let me know when you've located the sanding well [i.e., the one among the field's many that is producing silt] so we can take [mitigating] measures." The control room operator confirmed, "I'll set the lab assistants on it at once" and ended the call.

During this period, digital sand as a characteristic of the well flow became naturalized, and digital sand monitoring institutionalized, as a routine. As the routines and techniques investigating the organizational credibility of digital sand stabilized, however, NorthOil pushed further to improve how it used the digital sand-monitoring system to optimize daily production volumes, as explained below.

PERIOD 3 (EARLY TO MID-2000S): SAND, AN INTERACTIVE, ALGORITHMIC PHENOMENON

"So, look here, and see that we have a steep increase in the measured amount of sand flowing across this probe," Vinnie,[4] one of the onshore production engineers, said. Vinnie had been the on-call production engineer the night before, and the offshore control room operators had called him to investigate a sand alarm in one of the field's many wells. At the heart of Vinnie's retelling of the incident was a graph with plotted sand data in a time line. The first thing he had done upon receiving the phone call was to open the application that plots this graph. "I'm quite certain we have sand entering the well," he continued, "but then I look at the down-hole pressure here." He pointed to a green trend line plotted in the same coordinate system. "I realize that almost no fluids are streaming through the well. I would normally ask the control room operators to choke down [reduce the flow rate of the well] to prevent sand from damaging the production equipment. In this case, however, I am asking them to choke up. We are dangerously close to a shut-in pressure where sand will simply flow back down the pipeline."

Innocuous as this statement might have seemed, it bears witness to a significant shift in how NorthOil mitigates sand in the production system. The operational procedures for mitigating sand in the well fluids remained largely unchanged after introducing the digital sand-monitoring system in the onshore production organization in period 2 outlined above. There was a relatively clear and stable division of labor between off- and onshore tasks pertaining to sand management. The onshore production engineers were mainly used to confirm whether or not the sand alarm actually indicated sand. Once the onshore production engineers confirmed there was indeed sand in the well flow, the offshore control room operators would reduce the production rate from the well in question to limit fluid drag within the reservoir and, hopefully, the amount of sand swept along with the fluids being drained out of it. The problem is that with copious amounts of sand in the fluids, a loss of well flow velocity causes the sand to flow back and fill up the pipelines. But as NorthOil's leading expert on sand mitigation observed:

> Looking back at the data collected by the sand sensor system, the data was clear for those of us with knowledge of how sand producing wells behave. The production engineers at the operations center lacked this knowledge. They did not recognize the indicators before the pipeline was filled with sand and irreparable damage had been done.

The sand-monitoring software Vinnie was using came about as NorthOil initiated a large research project to improve operational communities' ability to handle sand incidents. Instead of representing sand as an absolute number that changes with each sand measurement, the digital sand system vendor made use of plotting sand data in a time series graph, as one software engineer with the vendor explained:

> The information was presented [in the user interface] in a way they could not relate to. It [the information] was just [presented as] a number, but what does that number mean? They needed to see trends and be aware of the system's limitations. They needed to consider factors that affected the measurements, but which were not sand related. So, if they had an alarm, they had to manually assess whether the alarm was an actual incident.

What appears to be a simple engineering trick for the vendor's lead software engineer, however, opened a window of opportunity for improved mitigation strategies. Sand in the production system has been a well-known problem within the international petroleum industry since the 1940s. By the 1970s, researchers within the earth sciences had formulated theories on the geomechanical properties of different reservoir rocks to explain the causes of sand in the well flow. Much of this knowledge remained within the scientific domain, and its use within operational settings remained limited. An important part of the research project was therefore to create a correspondence between the shapes of *trended* sand data in historical sand incidents with different theoretical explanations for sand entering the well flow, diffracted digital sand into different kinds of events. In this manner, the phenomenon of digital sand is made richer and more nuanced and has multiple triangulated sources. Digital sand, then, never came down to "capturing" sand by a singular, perfectly accurate sensor. On the contrary, it emerged over time by adding layers, nuances, and interdependencies to the algorithmic phenomenon of digital sand. A steep incline in the trended data, for instance, corresponds with a sand avalanche where the reservoir collapses around the well. Repeated spikes of sand data against a background of an otherwise low sand influx correspond with another explanation (e.g., *slugging*), and so on. Each explanation came with its own particular mitigation strategies. Using the newly available trended sand data to identify an avalanche threatening to fill the pipes with sand, for instance, would be mitigated by *increasing* the well flow velocity to quickly transport the sand out of the pipes, instead of the old mitigation strategy of choking down and thus *reducing* it.

This opened up activity between the sand data trends and existing geomechanical theory to determine the causes of sand influx. In the aftermath of a sand incident, a production engineer explained, "It is fairly easy, really. The only thing we can do is to increase or decrease the production flow."

This, however, is a truth with modifications, as production engineers would monitor how the reservoir reacted to changes in increased/decreased fluid flows. As the same engineer later explained: "I continue to monitor the sand graph after instructing the offshore control room operators to choke down, at the same time monitoring for pressure increase in the pipeline." The production engineers would also continue to monitor how other operating variables behaved. A pressure increase, for instance, would be an early warning signal that the sand was starting to flow back. If this was taking place, the production engineer would instruct the offshore control room to increase the well flow velocity, seeking to lift remaining sand out of the well. What we see is that the trend opened up a form of interactivity with the reservoir, in which the production engineers could monitor the effects of their mitigating strategies and adjust them accordingly.

The gradual development of a richer, interactive algorithmic phenomenon of sand feeding a deeper understanding of the individual wells' personalities gave rise to new mitigation strategies. While the basics of this geomechanical theory are comprehendible to nonspecialists, the time it takes to do such a diagnosis is still beyond the offshore control room operators' temporal horizon. By visualizing sand data in a time series, the digital sand-monitoring system speeds up the feedback loop between petroleum professionals' actions and their effects on the phenomena in the physical environment they seek to regulate. With this, a new, interactive way of working with sand in the well flow emerged as production planning (which is the production engineers' domain) and production control (which used to be the control room operators' sole domain) were conflated. Consequently, mitigation responsibility was transferred from the control room operators to the production engineers, who now made decisions on how to operate parts of the offshore plant.

The trend came at the fore of sand monitoring and mitigation practice, as production engineers naturalized the practices of poking and triangulating sand data outlined earlier. They kept correlating sand data with temperature and velocity measurements to determine if sand alarms were triggered by other factors. They also continued to work seamlessly with both physical and digital representations, although their reliance on physical inspections remained less relevant. Instead, they used well flow sampling. The well flow sample is a physical sample tapped from the topside pipes and then taken to the laboratory and separated to determine whether or not there is sand in the fluids. These activities remained important, as NorthOil retained its zero-sand tolerance policy to prevent environmental and human damage. However, an adjacent field had moved on to using digital sand to monitor for equipment erosion, allowing the production organization to produce even with sand in the well flow, urging NorthOil to push forward.

PERIOD 4 (MID-2000S AND ONWARD): SAND AS A MACHINE LEARNING–BASED PREDICTIVE MODEL

It is the daily coordination meeting between the production, operations, and maintenance engineers located in the onshore production facility and the off-shore process engineers and control room operators. The onshore engineers are gathered around a big conference table and linked with a similar-looking conference room at the offshore platform that also has several screens showing selected aspects of the status of the offshore platform (not unlike the bottom-most picture of figure 1.1). "Sand!" Howard, the meeting coordinator announces, as this is the item we have reached on the standardized agenda used for these meetings. "There was a sand incident in well A-6 last night," he continues. One of the attending maintenance engineers picks up on this, asking, "Do we need to inspect it?"

The "it" in question is the well's topside choke. Chokes are the valves that control fluid rates within the pipelines. Sand particles wear these valve openings down, degrading control over fluid rates. At worst, the valve casing is worn down, causing gas and fluid leaks. Chokes therefore tend to be replaced well before they wear out. However, replacing chokes has consequences. If choke erosion is suspected, the whole well is shut down as a precautionary strategy, causing a loss in productivity. Wells are otherwise shut down for scheduled maintenance only.

Everyone attending the meeting is of course well aware of this. They hesitate. For a second or two, the room goes quiet except for the whirring of computer-cooling fans. Ultimately, measurements and indications remain uncertain without physically inspecting the choke for erosion. Pete, the field's senior production engineer, breaks the silence, saying, "A-6 has a history of eating up its valves," which effectively makes the decision, knowing well the consequences in terms of productivity loss. Accordingly, Howard instructs the maintenance engineers to initiate an inspection of the choke, and the production engineers put A-6 on their list of nonproducing wells pending choke inspection.

The event with A-6 is later discussed over lunch. Kris, a senior research consultant with an international standardization organization, argues that "it is much like an egg of Columbus," referring to finding a simple solution for a seemingly intractable one, allegedly from when Christopher Columbus slightly crushed one side of an egg to make it stand still on a table. NorthOil, as with all oil operators on the Norwegian continental shelf, has until now vigorously pursued a zero-sand tolerance policy (due to the earlier explained risks to health, climate, and economic value). This policy, however, is being challenged. What Kris refers to as an egg of Columbus is a shift from preventing to managing sand erosion in the production system. What this implies is that wells, in situations like A-6, will not automatically be shut down whenever sand is detected (zero tolerance) to physically inspect the erosion of the choke. Instead, the well is to shut down only when the erosion of the choke is predicted to risk degrading process control. Hence, some sand is acceptable if "manageable"—that is, if the consequences of a loss of control are deemed noncritical. The crucial element in such a shift in sand policy is credible predictions of the erosion of chokes without actually (shutting down the well and) inspecting them physically. Together with researchers from NorthOil's R&D division, the sand-monitoring system vendor and the standardization organization developed an application to monitor the erosion state of chokes, pipelines, bends, and manifolds in the production system. Through tight collaboration they developed a predictive algorithm to simulate the erosion of valves and pipeline bends based on the offshore plant's real-time digital sand data. This allows a state to be simulated without being physically inspected.

In a first attempt, the predictive model in the simulation software was to rely directly on the IoT-generated sand data. As demonstrated earlier in this chapter (and, for exploration data, in chapters 4 and 5), this turned out to be a problematic assumption. A senior software engineer with the vendor admitted that when the "input data comes with a lot of uncertainties [and] the quality of the input data varies, the visualized output is basically meaningless."

After what was by all accounts considered a failed pilot test, NorthOil started another research project with the software vendor. Interestingly, and in the spirit of industrial science, NorthOil's research engineers did not go back to the drawing board to improve the reliability of the sand sensors. Instead, their knee-jerk reaction was to live with imperfection and instead find ways of dealing with it. Relying on the underlying predictive erosion model developed during the previous efforts, they decided to develop software functionality that, crucially, calibrated the simulations to another post hoc erosion measurement procedure already in place known as step-rate testing.

Step-rate testing is a calibration procedure used to determine the rate (volumes) of streaming from a single well into the topside processing plant at different degrees of choke valve opening (called steps). It is predominantly employed as part of the daily planning of production to ensure that the topside plant is used to its fullest processing capacity. The sheer force of the fluids streaming at high velocities wears the chokes down over time, increasing the valve opening the fluids stream through. The step-rate test is used to update the tabulation of choke valve opening and the rates to compensate for this wear and tear. Step-rate testing consumes invaluable production capacity as well as requires specialized equipment on the topside platform. It is therefore conducted only once or twice a year.

It was this existing calibration procedure that NorthOil's research engineers tapped into to calibrate the sand-based erosion predictions. Comparing the difference in measured fluid rates between the current and previous step-rate test, the engineers were able to develop erosion profiles to determine the current degree of erosion of a choke. The credibility of this regime was further strengthened when replacing chokes predicted to be worn out. Beaming with pride, the chief software engineer working on the condition-monitoring system stated, "We find that all chokes have eroded as predicted in all nine out of nine inspected."

Cultivating the organizational credibility of predicted choke erosion improved significantly when tethered to an existing calibration procedure (step-rate testing). However, it required substantial effort on the part of the production engineers to do this tethering to analyze the data to ensure calibration. Hence, it struggled to scale from a prototype to a routine tool. "Being able to drill down into the data and to actually correlate different data types is, of course, invaluable," a senior production engineer said before elaborating further:

> It gives us the chance of actually looking into the data and determine if we need to take action. As long as we monitor erosion on one, two or even a handful of wells, the tool is all we need. But on a field with 120 wells, that's another matter. We need some help to know which wells to pay attention to.

Again, the pragmatic instincts of industrial research came to the forefront. Rather than dismissing the prototype for predicting choke erosion due to excessive infrastructural work (see chapter 4), it was repurposed. Designed originally with the aim of generating predictions used to trigger mainte-nance interventions for individual wells, it was instead used to help onshore production engineers tasked with monitoring a portfolio of (hundreds of) wells. It was, accordingly, used as a screening or filtering device, filling the highly appreciated role of sorting the bulk of "unproblematic" wells from the smaller sample of wells warranting closer production engineers' scrutiny:

> We decided to use a predetermined sand rate—that is we feed the algorithm with an expected level of sand content for every well—to determine how ero-sion prone each well is. What we do is to simulate the consequences of having this fixed sand rate on the different wells on the field. Say we monitor one hundred wells. For eighty of these wells this sand rate will have no erosion con-sequence [i.e., it will not, within the set period of time result in erosion that is outside safe levels]. For these there is no problem. But for the remainder twenty wells erosion may be an issue, and the production engineers need to pay par-ticular attention to them. For these we have to ensure that the sand levels are so low as to not be a risk.

Sand monitoring as described above is used to prevent sand from incurring production loss while at the same time ensuring that the sand streaming through the system does not erode through pipeline bends and chokes. Pro-duction optimization is focused on optimizing the topside processing plant for maximal daily production volumes. Production optimization normally has a very short-term horizon. However, by using the sand-monitoring sys-tem's predictive ability to simulate the erosion potential of different produc-tion scenarios, production engineers started to make long-term production optimization decisions. Predicting the erosion proneness of pipeline bends and chokes translates into production optimization practice in that pro-duction engineers simulate different scenarios for running the production system and then determine how current production decisions for short-term production optimization affects production optimization in the long term. They look at which parts of the system can run with a lot of sand for a while before having to be shut down. Through simulations, short-term optimiza-tion decisions come to be entangled with long-term planning as engineers use the simulations to project the consequences of their short-term decisions on the long-term viability of the field.

CONCLUSION

Rather than a dichotomy, what is striking in the above analysis is the relative effortlessness with which engineers navigate among real physical sand, plotted graphs of indicated sand, and predictive simulation models of sand. For all practical purposes, they all fill the role of sand in NorthOil operators' daily work practices. Their attention is the epistemic concern of knowing sand, regardless of its many manifestations.

In this sense the discourse on the ontological status found both in STS (van Heur et al. 2013) and in debates on sociomateriality in organization studies misses the point (Orlikowski and Scott 2008). As Cecez-Kecmanovic et al. (2014) point out, "After all, how many practitioners are going to be able to make any sense of, never mind care about, whether we adopt a critical realist or an agential realist ontology, and so on?" (826). The engineers at NorthOil are not concerned with what sand—really—is.

If chapter 1 made the analytic argument for dismantling the physical/real versus digital/virtual dichotomy, the current chapter provides vivid empirical elaborations. The realness of digital representations of physical sand is a highly acquired quality, not in any meaningful way read off through the abstract principle of liquefaction. The evolving work practices of the different offshore- and onshore-based engineers demonstrate what over time went into the domestication or naturalization of digital representations of *sand* (Silverstone and Hirsch 1994), the seamless meshing of digital representations with existing practices. Digital representations, gradually and with effort, get woven into the moral economy of the engineers' everyday lifeworld.

Specifically, this chapter underscores the importance of practices of interactive poking—"playing," as Lehr and Ohm (2017) phrase it—with the representations of sand to gain familiarity and confidence. The ongoing quest to craft, not simply assume, organizational credibility is dominated by institutionalizing the necessary calibration, triangulation, and tethering to existing routines and procedures that physically access the chokes without adding additional cost.

7 POLITICS

written with Elena Parmiggiani

Science and technology—politics by other means.
—Latour (1993)

The oil and gas industry is increasingly controversial.[1] Climate agreements make abundantly clear that the fossil fuel paradigm needs to be dramatically reduced, if not dismantled altogether, to avoid the bleak scenarios associated with a rise of over two degrees Celsius in global warming from excessive carbon dioxide. As outlined in chapter 2, the offshore oil and gas industry has throughout its fifty years of history in Norway never been uncontroversial. The formative principle shaping Norwegian policies discussed in that chapter was that of a slow, not rushed, pace of development in order for Norwegian companies and associated institutions to learn and cultivate knowledge. Political opposition has mainly been motivated by concerns for Norway's substantial fishing industry and the environmental (Norwegian Ministry of Climate and Environment 2011). Historically, political deliberations in the Norwegian political environment around oil versus the environment adhere to, at least ostensibly, navigating an ideal of "knowledge-based," understood as neutral,[2] decision-making. Political majorities in various configurations in Parliament have been committed to a process that starts establishing a knowledge base (*kunnskapsgrunnlag* or *konsekvensutredning*) prior to making actual decisions. The regulatory approach of the Norwegian government regarding oil and gas activities ostensibly underscores the knowledge-based underpinning of granting operating permissions: "Official decisions that affect biological, geological and landscape diversity shall, as far as is reasonable, be based on scientific knowledge of the population status of species, the range and ecological status of habitat types, and the impacts of environmental pressures" (Norwegian Ministry of Climate and Environment 2009, sect. 8) Keeping certain areas off-limits for oil activities has been a key mechanism for balancing the inherently contradictory interests related to oil activities and environmental and fishing needs. Opening up areas in a phased manner allowed the kind of gradual learning embedded in Norwegian oil policies (see chapter 2). For instance, until 1979, Parliament imposed a ban on oil activities north of the 62nd parallel (roughly off the west coast of Norway) due to the perceived risks to safety and the environment, which was lifted when Parliament assessed that adequate experience and knowledge had been acquired.

The focus of this chapter is on a currently unfolding, intensively controversial question: whether to open up to oil and gas operations designated areas in the Arctic that are presently banned (see figure 1.4, with a map of the Norwegian continental shelf). At the core of this controversy are questions about what we know—and how we know it and with what degree of certainty—about potential consequences for the fishing industry and the environment as well as for employment and work in the north of Norway. For the purposes of this book, the controversy around Norwegian Arctic oil is illuminating. The political arguments are particularly sharp and clear. On the one hand, the case for the environment and fishing is compelling. The Arctic is a pristine environment and the only region in the world (together with the Antarctic) that could be described as relatively undisturbed by human footprints. Rich in flora and fauna, it harbors abundant fishing grounds. The Arctic is particularly vulnerable to environmental disasters such as oil spills because low sea temperatures radically reduce the speed at which oil naturally degrades. This is distinctly different from, for instance, BP's Deepwater Horizon oil spill in the warm waters of the Gulf of Mexico in 2010, where microbes contributed significantly to reducing the environmental impact of the oil spill because "the microbes chewed through the smaller, dispersed hydrocarbons (and the dispersants themselves) relatively quickly, helped by the fact that these molecules can dissolve in water" (Biello 2015).

Controversies feed off diverging assessments of the operational safety (e.g., harsh weather and ice) and the financial and environmental risks posed by oil and gas activities. The abandoned drilling campaign off Alaska in 2013 by Shell is illustrative (US Department of the Interior 2013). Ice movement severely hampered, and ultimately aborted, operations due to safety concerns. In addition, legal challenges are mounting to oppose Arctic oil that accuse oil and gas companies for negligent behavior and/or undercommunicating environmental risks. In much the same way as the campaigns against smoking proceeded, legal measures have been pursued (see, e.g., the Sabin Center for Climate Change Law at Columbia Law School). From this perspective, Arctic operations are "the new tobacco" (Salvesen 2016; see also Conway and Oreskes 2012). Legal challenges to Arctic oil thus transform environmental risks into financial ones for investors in oil companies by injecting uncertainties into the outcome of litigation.

On the other hand, the oil and gas lobby forcefully argues that an estimated 25 percent of unexplored oil and gas worldwide is in the Arctic (Bird et al. 2008). There is, accordingly, a significant potential, it argues, for employment and work in the northern parts of Norway, areas that historically have struggled economically. Compounding these arguments, the oil and gas lobby is motivated by the fact that most fields are "mature": the fields are nearing the tail end of their production life cycle (Thune et al. 2018). As Ryggvik (2009) observes, there is little appetite to engage proactively in the discussion of how and when to gradually phase out oil activities on the Norwegian continental shelf. Instead, there is strong lobbying for opening up new areas, such as those presently off-limits in the Arctic. Figure 7.1 illustrates the situation, showing the size of oil reserves relative to their initial, full reserves.[3] Newly discovered yet untapped fields are thus at 100 percent (*top of the vertical axis*), whereas nearly depleted fields are close to the horizontal axis (i.e., 0 percent).

The relevance, then, to this book is to analyze *how* marine environmental data pertaining to the Arctic are marshaled into institutionally and politically credible "facts." Consistent with the theme of digital oil, the Internet of Things (IoT) plays a crucial role in capturing data about the marine environment (chapters 1 and 4).

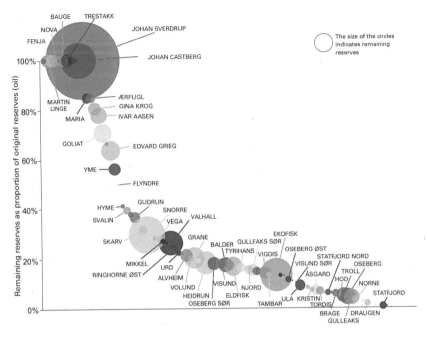

Figure 7.1
Remaining oil reserves relative to initial (100 percent) resources. The size of the circles indicates the size of the reserves.
Source: Reproduced by permission from the Norwegian Petroleum Directorate (2019).

THEORETICAL FRAMING: THE CRAFTING OF "FACTS"

The rich, open-ended, and ultimately qualitative phenomenon of the Arctic marine environment may be represented in numerous ways, depending on what aspect is highlighted. This taps into deep-seated science and technology studies sentiments of making the *contingent* nature of these choices visible through their social and political processes. Especially striking are the demonstrations employed to give voice to marginalized groups (Epstein 1996; Cipolla et al. 2017; Winner 1978) or particularly consequential areas of technoscience, such as nuclear energy (Hecht 2012), climate change (Edwards 2010), or development agencies (Jensen and Winthereik 2013).

Underlying any choice of re-presenting aspects of a phenomenon, here that of the Arctic marine environment, are assumptions about classification schemes shaping what and how data are captured. Infrastructure studies have had a particular commitment to unearth these invisible, taken-for-granted assumptions. Bowker and Star argue (2000) how the infrastructural inversion methodological principle allows us to deconstruct given categories—that is, demonstrate their contingent nature to reassemble them differently. As they point out through their study of classifications:

> As classification systems get ever more deeply embedded into working infrastructures, they risk getting black boxed and thence made both potent and invisible. By keeping the voices of classifiers and their constituents present, the system can retain maximum political flexibility. This includes the key ability to be able to change with changing natural, organizational, and political imperatives. (325)

Foucault (2005) famously made a similar point when quoting (Jorge Luis Borges's retelling of) the ancient Chinese encyclopedia that classified animals into fourteen categories that included "stray dogs" and "those drawn with a very fine camel brush."

Moving closer to the question of what and how to make selected aspects of the Arctic marine environment visible, it is necessary to admit that our knowledge of the marine environment, not only in the Arctic, is at best patchy. NASA's spacecraft Mars *Odyssey* mapped the surface of the Red Planet at a resolution of less than one hundred meters. As the futile search in 2014 and beyond for the wreckage of Malaysian Airlines flight MH370 in the Indian Ocean has made dramatically vivid, approximately 95 percent of the ocean—which corresponds to 70 percent of our own planet—remains unmapped. Satellite technology is useful for measuring water height, but it is not suitable for scanning the composition of marine ecosystems and the landscape of the seafloor because water blocks radio waves. Acknowledging that "we know more about the moon than the ocean floor" (Hauge et al. 2014), several governments have recently established large-scale, IoT-based sensing networks for generating new knowledge about marine ecosystems. For example, the Ocean Observatories Initiative in the US combines scientific platforms and distributed sensor networks that measure physical, chemical, geological, and biological parameters from the seafloor to the air-sea interface (Steinhardt and Jackson 2015). The MAREANO program developed by the Norwegian

Institute of Marine Research publishes detailed maps of the topography, geology, sediment composition, biodiversity (e.g., fish migration, coral reefs), and reported pollution of Norwegian seabed areas. Globally, the Long-Term Ecological Research Network is an international research enterprise consisting of at least forty networks that share data regarding different ecosystems and that document the effects of climate change (Karasti et al. 2006).

As argued theoretically in chapter 1 and demonstrated empirically in chapters 5 and 6, an overriding theme of this book is that what we know (about subsurface oil reservoirs, about sand, and about the Arctic marine environment) is invariably caught up in how we know it—through IoT-generated data with subsequent algorithmic manipulations. IoT-based facts about the marine environment are in every sense of the phrase constructed. Given that "the" marine environment, literally, is an open-ended phenomenon, what aspects of the environment get targeted and, equally important, for what purposes and for whom? Thus, this chapter analyzes the material and knowledge infrastructure underpinning the production of facts about the Arctic marine environment. The comprehensive IoT-based machinery for knowing the Arctic marine environment presented in the case below brings to the forefront the consequences, some intended and others not, of the material, political, and strategic design choices that go into the production of facts; the IoT-based crafting of politically contested facts. All facts, IoT-generated marine environmental ones included, need "good travel companions" (Morgan 2010) to travel; to conjure up robust facts requires a supportive machinery or infrastructure as discussed in chapter 1 and demonstrated empirically here.

THE CASE OF ENVIRONMENTAL MONITORING IN THE ARCTIC: DATAFICATION OF THE MARINE ENVIRONMENT

The empirical focus in this chapter is on two areas in the Arctic that are currently banned for drilling. One is in the High North of the Norwegian part of the Barents Sea, while the other is the Lofoten-Vesterålen-Senja area (dubbed Venus here; see figure 1.4, with a map of the Norwegian continental shelf that includes these two areas).

Both areas highlight the conflict with the fishing industry. The High North of the Barents Sea attracts a significant number of fishing vessels from the EU (European Union), Iceland, and Russia in addition to Norway. One part of the Barents Sea is already open for oil and gas activities. The previous ban for that section was lifted in 2013. As the lifting of previously banned areas north of the 62nd parallel demonstrated, present bans are anything but guarantees for continued, not to mention permanent, bans. The lifting of the ban for this part of the Barents Sea triggered outcries from environmental nongovernmental organizations (NGOs): "Our comment is that this is not knowledge management. Here we are in a situation where science adapted to politics, whereas it should have been vice versa" (Lorentzen 2015). Due to its location far north of Venus, a key conflict in this area evolves around the ice edge (i.e., the border of ice) because of the operational and environmental hazards, as Shell's experiences with Alaska made evident. The definition of the ice edge, however, is not a given but is highly controversial and debated. If the ice edge is defined further north than its present stipulation, as noted by one environmental NGO, "Norway would drill [for oil] farther north than anyone else . . . closer to the ice edge, further offshore, in extremely productive biologically and, hence, vulnerable areas . . . [with] increased likelihood and consequences of accidents" (Arnadottir 2016).

The relationship with Russia is a central part of the larger geopolitical backdrop to oil and gas operations in the High North of the Barents Sea. The northern part of the Barents Sea was for several decades de facto banned from oil operations simply because the borderline dispute between Norway and neighboring Russia was unresolved. Negotiations started in 1970, and it was only forty (!) years later, in September 2010 during the (relative) thaw following the Cold War, that the border dispute was formally settled. The oil industry's lobbying for access in the High North of the Barents Sea resonates deeply with the present right-wing government, as expressed by then prime minister Solberg in 2018:

> We should not be overly passive towards Russia. We have to map the seabed to exercise our rights to the rich resources that the Barents Sea has to offer. We risk ending up in a weak negotiating position in the case of any potential [oil

and gas] fields cutting across the [Norwegian-Russian] border—if we remain passive without demonstrating our interests in these areas. (Gjerstad 2018)

With Norway a long-time NATO member, the relationship with Russia is contentious in Norwegian foreign affairs, although practical collaboration on the ground is thriving and has a long history. This was sedimented when Russia liberated the northern parts of Norway from the Nazi occupation in the closing stages of World War II and subsequently pulled back to the original borders, in contrast to Russia's campaign into Eastern Europe.

Venus is literally among the richest fishing grounds on earth. The warm water fed by the Gulf Stream from the Gulf of Mexico mixes with the cold waters from the Arctic to generate a fertile habitat for plankton, which subsequently form the basis for larger species. The migratory paths of pelagic cod mean that after living most of the year in the Barents Sea, cod come into the Venus area for spawning yearly. The Venus area thus competes only with Newfoundland for the richest cod-fishing grounds for Atlantic cod. In addition, herring, pollock, mackerel, and halibut are also commercially important. Venus has since before the Viking Age been a crucially important source of subsistence for large parts of the population, providing an independent source supplementing the always variable agricultural outcome in the Arctic. Venus has a rich economic and cultural history in addition to attracting a significant number of tourists due to its stunning scenery. Accommodating oil activities as part of the social, cultural, and economic lifeblood of Venus thus generates heated debate and emotion.

Responding to the relative paucity of a knowledge base, as pointed out by the Norwegian Ministry of Climate and Environment (2011) conceding that "information on the geology and petroleum resources is more limited, and the seabed had not been mapped in as much detail [as in other parts of the Barents Sea]" (10), there are initiatives to increase the number of seismic surveys (see chapter 3). Seismic surveying, according to the oil and gas industry, is a noninterventionist activity that avoids the risks to the environment that later stages of oil operations bring. What constitutes an intervention in the marine environment, however, is contested. As environmentalists and the fishing industries have noted, seismic surveys disturb whales and other mammals, with unknown long-term effects (Folkeaksjonen 2017). As depicted in figure 7.2, a considerable number of seismic surveys have been completed in the Lofoten-Vesterålen-Senja (Venus) area.

Moreover, the same seismic "data" can be interpreted differently by various groups. For oil operators, they indicate the "viability" of their activities, whereas for the fishing industry, they indicate the "vulnerability" of fish stock (Blanchard et al. 2014).

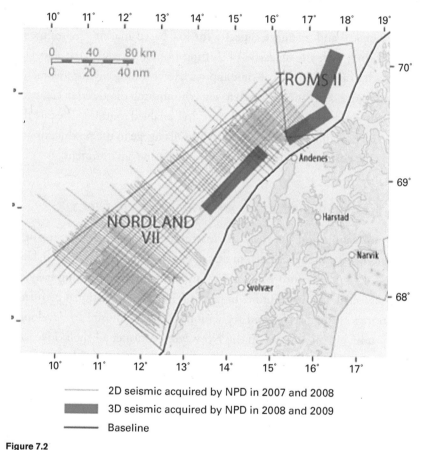

——— 2D seismic acquired by NPD in 2007 and 2008
▬▬ 3D seismic acquired by NPD in 2008 and 2009
——— Baseline

Figure 7.2

Map of existing seismic surveys completed in the Lofoten-Vesterålen-Senja area. Each straight gray line indicates a seismic survey, and those boxed in purple represent 3D, not just 2D, seismic. *Source:* Reproduced by permission from the Norwegian Petroleum Directorate.

In the debate on Arctic oil, there is a wide spectrum of stakeholders. This includes governmental and public agencies, industrial lobby groups, corporate interests, labor unions, and environmental activists. These stakeholders actively contribute to generating, supporting, and legitimizing knowledge of the marine environment in the Arctic. Shifting from mere consumers, oil companies, NorthOil included, are increasingly engaged in the production of data about the environment. Several stakeholders, notably environmental activists and fishermen, challenge the knowledge produced by oil companies (Lamers et al. 2016). Since the early 2000s, NorthOil has strengthened its technological and scientific capacity for IoT-based marine environmental monitoring, with the objective of shifting from a corrective, ex post facto approach to a preventive, real-time approach for assessing environmental risk. This mimics exactly the approach to sand monitoring discussed in chapter 6, which similarly moved toward real-time, IoT-enabled routines. The case thus illustrates how oil operators such as NorthOil engage in the production, not merely the consumption, of facts about the marine environment.

IOT: A VIABLE ALTERNATIVE? THE VENUS PROJECT

The motivation and approach to marine environmental monitoring at NorthOil was strikingly similar to that of sand monitoring discussed in chapter 6. Like sand monitoring, marine environmental monitoring has been an established, manual routine. In fact, it is imposed by national petroleum authorities through rules and regulations as a prerequisite to developing and producing oil and gas on the Norwegian continental shelf. Like sand monitoring, the traditional routine for marine environmental monitoring is to take physical samples of selected elements of the marine environment, which are then shipped to onshore laboratories for tests, the results of which get stored in databases for subsequent analysis. Like for sand, the routine is a labor-intensive endeavor, using hired contractors with special vessels to conduct the sample taking, as well as time-consuming, with the whole routine typically taking months to complete. For the same reasons as noted in chapter 6, it was targeted for cost cutting through digitalization, notably delegating to sensors the tasks of "sensing" the marine environment. Unlike sand monitoring, however, recent changes in regulations in Norway provide additional motivation for a less time-consuming routine, underscoring that "sufficient information shall be obtained to ensure that all pollution caused by own activities is detected, mapped, assessed and notified, so that necessary measures can be implemented" (Petroleumstilsynet [Petroleum Safety Authority Norway]; Ptil 2016) that lend credibility to an IoT-based, *real-time* monitoring of the marine environment.

Although explicitly specifying the water column, the sedimentation, and the seafloor fauna (Norwegian Climate and Pollution Agency 2011), government regulation is lacking in detail as to what needs to be covered and by which method. As a result, there is significant room for the oil operators to devise their own routines. NorthOil focuses on environmentally sensitive flora and fauna, with most common parameters in the water column consisting of oceanographic data (pressure, temperature, and salinity), the direction and speed of water currents, turbidity (instantaneous concentration of particles), sedimentation (long-term accumulation of particles on the seabed), and visual inspection through pictures and videos.

In an early and explorative effort to test the technical and practical viability of using an IoT-network to "capture" aspects of the marine environment, the low-key Venus (anonymized acronym) project was created in 2005 in the Venus area to establish an ocean observatory (Venus for short). Inspired by international examples including those referred to above, a small group of environmental advisers and technologists set out to test the viability of IoT in this context. Starting modestly, the Venus project experimented with several types of sensors to explore what, if any, of the Arctic marine environment they were able to target. Specifically, the project explored a variety of IoT configurations to measure water quality, acoustic devices to detect the concentration of biomass, and subsea cameras to capture photos of a rare species of cold-water coral reef, of which the world's densest population is located in Norwegian waters. The data sets were stored on hard disks for later retrieval. Being a small project, Venus operated within a limited budget. It accordingly had tight constraints on the type and quality of its sensing equipment. This remained the situation for several years while Venus collected data without much attention internally.

This changed decisively in 2010–2011 when the Arctic was elevated to a strategic area vital to NorthOil's long-term competitiveness. The Arctic thus rose to the forefront of strategic attention, spearheaded by the high-profile corporate Arctic Program. The importance of the Arctic was obviously motivated by the considerable estimated hydrocarbon reserves there. The *timing*, however, was reciprocally fed by—and feeding into—a broad, ambitious, and strong push by the Norwegian government, with due attention to Russia. In a series of policy white papers, followed by funded initiatives into industrial collaboration and cultural exchange projects, "the potential oil and gas resources in the north has been identified as the primary motivation behind the new policy in the north" (Gjerde and Fjæstad 2013). Hence, NorthOil's strategic Arctic Program mirrored, and fed into, the government's strong promotion of the Arctic, resulting in the vitally significant settling of the forty-year dispute with Russia over the offshore border between the two countries.

Against this backdrop, the Venus project received renewed corporate attention. Moreover, the Arctic Program saw Venus as promoting NorthOil's image of an environmentally friendly corporation. One NorthOil environmental adviser candidly explained the utility of the Venus project:

> There is of course one reason why we are doing this: it is to gather background data for potential future [oil and gas] operations. . . . We don't know. But in the meantime, we have this observatory, and we are going to use it for testing both software and hardware technologies.

After the recognition from the Arctic Program, additional funding followed to turn the Venus project into a permanent IoT-enabled marine-monitoring station approximately twenty kilometers off the coast of northern Norway, with a fiber-optic cable connected to an onshore data center. In addition to devices to measure the pressure, temperature, salinity, and cloudiness of the waters, acoustic sensors are used to detect moving biomass in the water column. Sensors are installed on a semipyramidal metallic structure weighing 400 kilograms, connected to a 1.8-meter satellite crane via a 50-meter subsea cable. A camera with a flash is situated on the crane, potentially to assess a coral reef previously identified by environmental experts (see figure 7.3). Completed

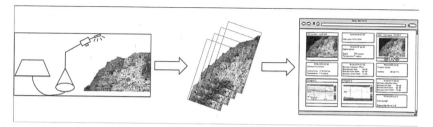

Figure 7.3
Outline of the sea observatory, with a camera installed on a crane on the seafloor in Venus to detect marine resources in the proximity of a coral reef (*red rectangles, figure on the left*). Photographs are taken every thirty minutes, transferred via a fiber-optic cable (*center*), and visualized via a web portal in real time (*right*).
Source: Reproduced by permission from the MAREANO/Institute of Marine Research, Norway. Art design by Elena Parmiggiani

in 2013, it provided the first openly available, IoT-based, real-time marine environmental data sets in Norway.

In contrast to the case of sand monitoring, where you know what you are looking for but not how, the Venus project grappled with what phenomena of the marine environment to target. As a result, the project went through several rounds of exploring, experimenting with, and improvising what to capture and how, governed by technical, material, and economic constraints. To illustrate, consider the camera on the crane (figure 7.3), which was included even though it did not immediately support any designated measurement method beyond its potential to capture corals. Hence, it was underutilized. Through internal discussions in the project group, the researchers came up with the idea to repurpose the camera to capture a relevant environmental parameter otherwise difficult to grasp, that of measuring the sedimentation. The idea, as one environmental adviser explained, was to take pictures of a sediment trap against a contrast of black and white every half hour to measure the amount of sediments accumulated during that period:

> We have sediment traps [i.e., tubes that physically trap sediments in the water] but . . . you don't have a way to electronically transferring it; you just gather sediment in a tube and take it off. . . . But if you connect a camera to it . . . that's new; it's something nobody has used!

The fishing industry was and still is a key antagonist of the oil and gas lobby, pushing for the Arctic in general and the area of Venus in particular. NorthOil recognized the political importance of forging alliances, when possible, with the fishing industry. The oil and gas lobby is eager to stress that collaborative relations with the fishing industry do exist. As one proponent underscored, "Fishermen have unique local knowledge and can mobilize at short notice. Their recruitment will provide a further strengthening of the oil spill preparedness organization near shore" (Norsk olje & gass 2017). Similarly, another representative of the oil and gas lobby stated, "We have already solved many challenges together with the fisheries. Among other things, we have done a lot together to find a time window for seismic shootings" (Fenstad and Hagen 2017).

The Arctic Program through the Venus project tried a variety of ways to enroll the fishermen. The fiber-connected, IoT-based Venus Ocean Observatory needed a high-speed connection from the onshore data center, not only on the seabed. In collaboration with local fishermen, NorthOil thus decided to finance a fiber-optic internet connection to the fishing village near the data center, thus creating spill-over benefits for the village beyond the Venus project's immediate demands.

With the funding from the Arctic Program, the Venus project developed a web portal to store, manage, and visualize real-time measurements (see figure 7.4). The Venus portal was published under a Creative Commons license and is openly accessible online. The data sets are owned by NorthOil but may be downloaded for further use or publication, assuming due acknowledgment. Thus, NorthOil sought to make the open Venus portal relevant and useful to the fishermen. As one NorthOil environmental adviser enthusiastically put it, "We can see fish, so imagine that the local fishermen can go in there and look: 'Is there a point in going to the sea today? Do [the fish] stay at home?'"

A workshop was organized in November 2013 to present the Venus portal to a community of fishermen. The feedback from the fishermen, rather unexpectedly, was quite positive according to our NorthOil informant present. A representative of the fishing community commented that he wished there were more such observatories in the area because they would be useful for predicting the amount of fish available to catch each day.

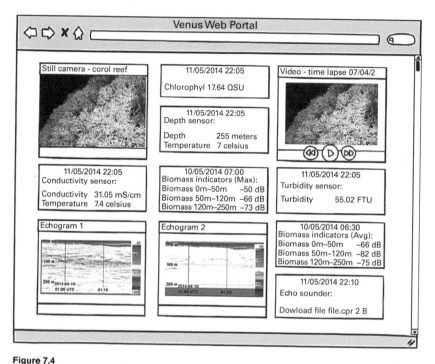

Figure 7.4

Outline of the web portal developed as part of the Venus project.

Source: Reproduced by permission from the MAREANO/Institute of Marine Research, Norway. Art design by Elena Parmiggiani.

MAKING ENVIRONMENTAL DATA RELEVANT TO DAILY OPERATIONS: THE ENVIROTIME PROJECT

Independent of the Venus project, another NorthOil initiative for environmental monitoring was in operation in 2011–2014. The EnviroTime project (a pseudonym) was a corporate initiative for environmental monitoring, headed by NorthOil in collaboration with a consortium of industrial partners covering key areas of expertise such as subsea technologies, information technology integration, and environmental risk assessment.

In contrast to the Venus project, which essentially was "playing" with the environmental data (Lehr and Ohm 2017) to explore what was feasible and useful with IoT networks, the EnviroTime project had a clear, strategic aim. It's goal was, like sand monitoring, to transform the marine environmental-monitoring routines from manual and time-consuming to datafied and real time. In addition, however, it aimed to transform other parts of daily operations (see chapter 3), such as drilling, production, and maintenance. In this manner, marine environmental monitoring would not be a self-contained routine but would shape and influence other parts of daily operations; EnviroTime pushed for an expansive role for marine environmental monitoring.

Expecting stricter safety and environmental regulations for oil operations in the Arctic, NorthOil was eager to preemptively change its marine environment-monitoring routines. Specifically, the EnviroTime project was to establish the machinery for operational routines through a "platform . . . to monitor and analyse the environment in parallel with daily operations in order to protect sensitive areas and to minimize the risk of potential negative impact on the environment" (Internal documentation, NorthOil).

Exactly as for the Venus project, the immediate challenge facing EnviroTime was *what* and *how* to capture the unbounded variety of the marine environment with IoT. Like Venus, the selection of aspects of the phenomenon to target was regulated by practical and technical constraints and possibilities with the IoT network. However, with the more expansive ambition of the EnviroTime project, there was also a particular focus on selecting aspects that were, or could be, relevant for politically charged issues related to Arctic operations—notably, investigating commercially interesting fish as well as protecting sensitive cold-water coral reefs.

The range of physical objects, qualities, and processes measurable to sensors is wide and rapidly expanding (Singh et al. 2014). In marine environmental monitoring, the list of physical objects, qualities, and processes targeted for sensor-based monitoring comprises water currents, pressure, temperature, conductivity, turbidity, oxygen, carbon dioxide, oil-in-water emulsions, methane, chlorophyll fluorescence, topography, and benthic communities. In short, sensors capture an expanding richness of physical objects, qualities, and processes. The EnviroTime project thus explored broadly the possibility of feeding a digitally mediated, real-time presence of the marine environment into operational work routines. Several threads were pursued, two of which are elaborated here.

The first thread starts from the observation that the Norwegian continental shelf is home to the world's largest number of cold-water coral reefs (Fosså et al. 2002). *Lophelia pertusa* is a species of cold-water coral living at a depth of forty to four hundred meters. Researchers estimate that 30–50 percent of the reefs have been damaged since the 1980s by the extensive bottom-trawling activity of the fishery industry, one of Norway's leading industrial sectors. There is, accordingly, political attention and concern to ensure that the oil industry does not create additional damage. *Lophelia* is vulnerable and was in 2003 included in the list of threatened species by the North Atlantic OSPAR Commission (2008) within the EU.

Most acutely, the operation of drilling a well is potentially threatening to *Lophelia* (see chapter 3). At five to forty inches in diameter, an oil well is not very wide. However, when drilled thousands of meters into the subsurface, they generate a considerable amount of drill cuts (cuts and particles from the rocks being drilled through) that may damage the surrounding marine environment. Compounding the environmental risk, it is necessary to lubricate the drill string with chemically modified drill mud that seeps out into the sea. Water currents may transport the pollution over larger distances. As a well is later secured to start producing, small, hardly noticeable leakages can still occur. Spills and leakages, environmental agencies argue, have more severe impact given the vulnerable marine environment of the Arctic. Against this backdrop, one representative from the drilling and well department at NorthOil declared at an internal meeting that "[the EnviroTime project] must produce reliable and trustworthy data about the environmental impact of the drilling operations to ensure it be taken into consideration as an operational modifier." This led to targeting *Lophelia* specifically for environmental monitoring. As one environmental adviser recalled, "We needed to do something . . . to find out whether these guys [i.e., the corals] are sensitive or not for the [drilling] discharges."

In response, the EnviroTime project developed a tool that predicted the distribution of the drill cuts using IoT-based observational data together with simulation models of the water current transportation. The tool is a map with a real-time update that predicts present and future risks for the coral structures. As the drilling activity begins, the tool provides an updated picture of potential changes in the impact of the drilling discharges—for example, from a change of the water currents. In this way, a vulnerable area of *Lophelia* was saved by shifting the location of the drilling further away and drilling horizontally into the subsurface to reach the suspected hydrocarbon reservoir.

CALIBRATING AND TRIANGULATING MARINE ENVIRONMENT DATA: AN ALLIANCE WITH VENUS

A recurring theme in this book is how to craft credibility into digital oil data through calibration, triangulation, and quality control. Chapters 4–6 all provide vivid illustrations of the different ways this is conducted. The EnviroTime project faced similar challenges in giving sufficient credibility, hence organizational significance and consequence, to marine environmental data.

In a second endeavor, EnviroTime, encouraged by the Arctic Program, forged an alliance with the Venus project. The thought was that the Venus data, accumulated over half a decade at this point, could be used by the EnviroTime project as a baseline against which to calibrate and triangulate its own data. To illustrate, consider the specific construct of biomass. Biomass, in its simplest form, is a construct to measure the presence of commercially interesting fish, the principal political concern in the Venus area. The construct, literally, is an algorithmic phenomenon (see chapter 1). As elaborated below, it is inferred in part from IoT measurements and in part from synthetic (i.e., model-generated) data, then algorithmically manipulated and presented.

However, to fix the construct of biomass meant navigating around both practical and political constraints. With fishing being Norway's second-largest export industry (after oil and gas), it was important for the IoT to "see" commercially interesting fish, such as cod, herring, mackerel, and pollock. One of the companies involved in EnviroTime, a market leader of subsea sensor technologies in Scandinavia, proposed to deploy state-of-the art sensors that send an acoustic wave and measure the echo returned when the wave hits a target, such as a fish (like a radar). The acoustic devices in EnviroTime were placed on the seafloor at a depth of a few hundred meters.

A first design dilemma was the trade-off between the reach and granularity of the acoustic sensors. The boundaries are governed by laws of physics dictating that, keeping energy consumption constant, you can either cover a wide cone in the water column with long wavelengths (i.e., low frequencies) or a much narrower cone with shorter wavelengths (i.e., higher frequencies) but not both at the same time. The operational relevance of the trade-off around determining the frequency/wavelengths to the context of environmental monitoring in NorthOil was that while fish were detectable with long wavelengths, small fish eggs and larvae from spawning were not. A reasonable assumption was that these organisms are more sensitive to pollution because they cannot react and swim away like fish can. For the long-term goal of positioning NorthOil vis-à-vis areas presently banned from oil and gas operations, monitoring fish eggs and larvae was, accordingly, particularly important. In addition, due to the budget constraints of the project, relatively inexpensive acoustic sensors had to be procured—sensors with shorter reach (from lower energy consumption). The implication for the project was that the uppermost parts of the water column, including the surface, were difficult to reach from a depth of a couple of hundred meters. Thus, the acoustic sensors were unable to detect fish larvae and eggs floating close to the surface outside the response range of the sensors, as one environmental adviser at NorthOil pointed out: "We can't, for instance, [measure] larvae and eggs in the upper water masses because this [sensor network] is on the bottom [at 250 meters down] and has a reach of about 50 meters." Although a fish was large enough to be detectable for the wavelengths associated with an acoustic cone of reasonable width, you still could not "see" all the fish. What was most clearly detectable from the acoustic signal was the fish's swim bladder, not the fish itself. The swim bladder is a gas-filled internal organ that fish employ to regulate depth when swimming. It functions as a resonating chamber for the acoustic sensors, and therefore fish with swim bladders are more easily detected than fish without. As one of the technology partners in the EnviroTime project explained:

> A big fish or a big swim bladder will return a bigger signal than a smaller one. . . . Species like the mackerel, which don't have a swim bladder, will return

a very small signal. Perhaps that's why we have come up with species with a swim bladder in this project.

What this adds up to is that, given the physical parameters of the acoustic sound sensors employed in the EnviroTime project, fish without a swim bladder (e.g., mackerel) and eggs were invisible. The lack of data about larvae and eggs threatened EnviroTime's ambition and political significance. It implied a lack of data about future generations of fish affected by a possible oil spill.

In response, the EnviroTime project transformed its marine environmental monitoring from a purely IoT-based, data-driven ocean observatory into an algorithmic phenomenon. More specifically, the missing data were compensated for by *synthetically* generating them from predictive, theoretical models and supplementing them with environmental data from the Venus project (Edwards et al. 2011). The Venus project, EnviroTime members hoped, could provide the much-needed historical data sets to feed a predictive model based on algorithms from one of the partners to generate missing data about particle dispersion in EnviroTime.

Using biomass as an indicator for marine environmental monitoring assumed having established baseline conditions. Here, the EnviroTime project used the Venus project data (actual and model generated). To make biomass organizationally real, it had to tie in with NorthOil's institutional practices and vocabulary of risk assessment familiar to professional groups—for example, the drillers and production engineers involved in daily operations. Existing practices relied on a semaphore-like visualization of risk. In an effort to tap into existing practices and symbols, EnviroTime decided to mimic such a visualization of risk and worked on methods to visualize biomass concentration in the water.

Adapting an approach previously devised by the Norwegian Directorate of the Environment,[4] the EnviroTime team formulated the environmental value, a number expressed in decibels that summarizes the hourly concentrations of biomass in large cubes of the water column (e.g., from depths of fifty to one hundred meters). The environmental value is obtained by collapsing multiple original sections scanned by the acoustic sensors into a single section.

Drawing on the visualization of risk from control rooms, the Enviro-Time participants developed a categorization structure based on five colors to classify the amount of biomass in the entire water column, summing up all the cubes in the water column. For example, the highest concentration was coded in red and indicated the highest-risk probability for the marine environment, which was to indicate a halt in operations.

THE POLITICS OF MODELING: THE ICE EDGE

NorthOil was during this period strenuously lobbying for lifting the then imposed ban on oil and gas activities in the High North in the Barents Sea, which is significantly farther north than the Venus area.[5] A particularly challenging issue when pushing north, readily appreciated by NorthOil's Arctic Program, was the ice edge. Given the operational and environmental risks associated with ice, NorthOil's capacity for marine environmental monitoring had to be credible not only in the Venus area but also further north in the vicinity of the ice edge. In other words, the Venus project needed to generalize. As the head of the Arctic Program explained, using the proverbial Norwegian "potato" as a metaphor, the ice edge is a proxy of generalizing as follows:

> The Arctic Program wants to position [NorthOil] in the north, and the ice edge, in particular, is a very good potato: they need food to position [NorthOil] in the North. . . . What is interesting for the Arctic Program is the issue of the ice edge, what resources are there, what can be visualized. There are many political aspects involved.

The contested area regarding the ice edge is the High North of the Barents Sea toward areas to the south of Svalbard (see figure 1.4). The central controversy is misleadingly simple to express: Where is the ice? Given the well-known operational hazards connected with oil operations in the presence of ice (US Department of the Interior 2013), there is broad consensus about the necessity to avoid ice.

However, the extent of the ice south of Svalbard varies significantly across seasons and time periods. The role of man-made climate change is particularly vivid at Svalbard, with its permafrost environment, as "nowhere on the globe are the effects of climate change more pronounced" (Meteorologisk Institutt 2019).

The political controversy about the ice edge pans out in public discourse and policy documents as, seemingly, a scientific controversy. If there ever was an instance of the slogan about politics by other means, the ice edge would be a perfect illustration. Table 7.1 outlines the ongoing debate, with four distinct positions—expressed purely as alternative *methods* and *data sets* for computing the ice edge—tied to political interests. Controversies over data vary from presently observed ice to data from the last thirty or fifty years, while controversies over methods vary from those areas with a minimum of 30 percent ice in the month of April to areas covered with ice anytime during the last thirty years. Figure 7.5 depicts the consequences of each of the four positions.

Table 7.1
Four different methods of defining the ice edge, each relying on different data sets, as evident in discussions in 2017. See Rommetveit et al. 2017.

Principal proponents	Definition	Implications
Minister of oil and energy	The observed ice edge—that is, where the ice edge at any time is present (*dashed line*, map in figure 7.5).	Ice edge will vary dynamically on a daily basis. Given existing observations, the ice edge will be shifted significantly to the north of the current definition.
Government's official position	Those areas with a minimum of 30 percent ice in April during the time period 1985–2014. Concentration of ice needs to exceed 15 percent (*green line*, map in figure 7.5).	Ice edge will be shifted to the north, albeit not as far as the above proposal.
Parliamentary majority	Areas with a minimum of 30 percent ice in April during the time period 1967–2014. Concentration of ice needs to exceed 15 percent (*red line*, map in figure 7.5).	This is the present definition and the basis of current permits.
Environmental activities	Areas covered at any time during the last thirty years (*yellow line*, map in figure 7.5).	This will shift the ice edge significantly to the south. It will ban some of the oil fields currently operating in the northernmost part of the Barents Sea.

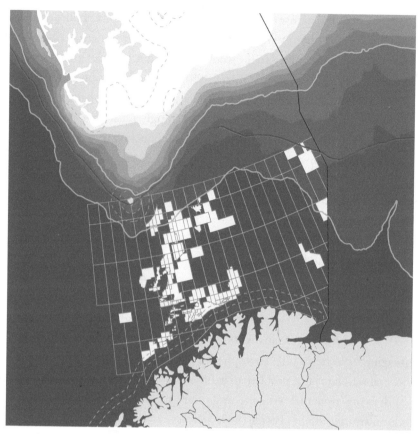

Figure 7.5

Map of the four different ice edge definitions listed in table 7.1.

Source: Reproduced by permission from Norsk olje og gass (see Johnsen 2020).

CONCLUSION

The phenomenon of the marine environment is inherently open ended, as argued by Knorr Cetina (2001; see the discussion in chapter 5), extending "infinitely" while the properties of the phenomenon are evolving. Thus, where does it start and end, and central to this book, how do we gain knowledge of the environment? The vastness of the phenomenon, compounded by the inaccessibility of data about it, relegates qualitative assessments and tactile or practice-based approaches to knowing to the margins. This lack of or the impractical nature of immediate access to the empirical phenomenon under scrutiny is hardly unique to the marine environment. Other cases include but are not limited to the study of subatomic particles in high-energy physics accelerators (Knorr Cetina 1999) or the exploration of the planet Mars (Vertesi 2015). Edwards (2010), in a study of the open-ended phenomenon of global climate change, compellingly illustrates instrumented knowing. Edwards reconstructs the historical trajectory of the *vast machine*, a global, interconnected network of instruments, sensors, and satellites that gather data sets about the atmosphere, the oceans, and other relevant natural systems to generate and validate climate models. This vast machine transforms remote observations into accepted knowledge by enrolling "scientific expertise, technological systems, political influence, economic interests, mass media, and cultural reception" (8).

The reliance on instrumented, necessarily quantified modes of knowing implies a reduced role for the qualitative and tactile (although these qualities, partly, get reintroduced through bodily gestures, such as Alac's (2011) study of physicians working with functional magnetic resonance imaging or Vertesi's (2012) emphasis on an embodied knowing of Mars). Several scholars have problematized the flatness of quantified knowledge. In drawing a parallel with the way quantification pervades political and administrative affairs, Porter (1996) shows how quantification in the science of nature has evolved and has developed a strategy of impersonality in response to outside pressures. As such, quantification is a powerful means to communicate knowledge outside of a local setting and to make it global by coordinating activities and settling controversies: "It implies nothing about truth to nature" (Porter (1996, xi). Indeed, it has more to do with flattening the qualities of knowing practices into manageable categories.

Such a collapse of different qualities into common metrics—commensuration—transforms quality into quantity (Espeland and Stevens 1998). It is an unavoidable condition when conducting remote monitoring of the environment. By means of commensuration, disparate information can be compared and fed into mechanized decision-making, and relations between attributes can be revealed. The social and material/technical dimensions of knowledge production (facts) can thus coevolve. For instance, in

Kallinikos and Tempini's (2014) study on the production of health facts using a social media platform, they discuss how the quantified traces of user behavior are constructed dynamically and algorithmically through the coevolution of the platform and selected aspects of user interactions. An example is how self-reported symptoms among patients, traced by the platform over time, are given visibility and recognition by the physicians and medical researchers associated with the platform.

Commensuration has strong political connotations. In the age of datafication (Lycett 2013), systematically adopting metrics and indicators to measure everything from societies to nature tends to frame political debates into technical vocabulary. This shifts corporate modes of auditing and governance into the public sphere (Merry 2009). In other words, by making the nonmeasurable measurable (e.g., from an open ecosystem to a quantified degree of environmental risk), commensuration flattens out complex political and ethical concerns that might otherwise lead to conflict and confusion (Espeland and Stevens 1998). Attending to the way systems flatten out qualities "enables us to appreciate the extent to which commensuration constructs what it measures" (329).

General arguments against the quantification of quality are also amply present in the specific case of the marine environment. Marine policy research has warned against missing out on the uncertainties embedded in the process of defining and quantifying knowledge of marine life (Blanchard et al. 2014). Ecosystems are never unambiguously given, but the facts that constitute our perception of what counts as environment and environmental risk are constructed by means of the scope of the instrumentation and risk-assessment methods (such as by disregarding noncommercial fish species), the methodological choices made (such as by relying on worst-case scenarios to monitor pollution rather than during routine fishing or petroleum activities), and the manner in which results are presented (such as by traditional risk-assessment methods; Hauge et al. 2014). The categorization methods used to classify the environment must be made compatible with corporate or administrative governance, which is frequently driven by financial motivations (Knol 2013).

III IMPLICATIONS

8 CONCLUSION

The theme of the digital and digitalization dominates this book, as signaled by the title. The notion of digitalization has entered everyday vocabulary in unprecedented ways, thus punctuating the normally all-too-effective separation of academic from public discourse. Fear and awe are evoked in equal measure in the public media covering all spheres of everyday life, be it in your home with AI working as "your own butler" (D. Brown 2021) to perform chores, at work where "robots are coming" (Deming 2020), in places caring for the elderly with chatbots (Mateescu and Eubanks 2021), or in museums featuring artificial intelligence (AI)–based artists (M. Brown 2021). This ongoing public debate demonstrates, at a minimum, how digitalization is experienced as present and relevant in ways it has not always been. In this sense the digital is experienced as phenomenologically real in our everyday lifeworld (Boellstorff 2016). Coined decades ago by cultural theorists, digitalization is *domesticated* into the moral economy of everyday life (Silverstone and Hirsch 1994).

The broad interest into digitalization, however, comes with an ambiguity: the discourses, spanning from the public to academic, are made possible because of, not despite, ambiguity or underspecification of the concept (Swanson and Ramiller 1997). This motivates a principal concern of this book: What *is* digitalization and what does it *entail?* Against the backdrop of practices of knowing in the Industrial Internet of Things (IoT), what is the emerging picture of digitalization, and what are the contours of digitalization offered?

In this book, digitalization is conceptualized as *efforts to quantify the qualitative*. There are important caveats that need to be addressed, however, before we get into what such a perspective amounts to.

First, the qualifier "efforts" (of quantification) is essential. As demonstrated in the empirical chapters of part II of this book, the efforts to quantify are precisely that, efforts. Quantification may well be the ambition, but its realization is littered with setbacks, resistance, infrastructural work, or outright dismissal. Hence, the resulting level and type of digitalization is an amalgam of the arc of the ambition with counterreactions.

Second, efforts to quantify, clearly, are not new. On the contrary, historic accounts of processes of quantification analyze the emergence of modernity and its institutions through this lens (Crosby 1997; Espeland and Stevens 1998; Igo 2007; Merry 2009; Porter 1996). For instance, Didier (2020) gives a rich, historic analysis of the collective, distributed, and heterogeneous practices—recruiting statisticians, editing and commenting on questionnaires returned from respondents, refining mathematical notions of random sampling and sample frames—involved in crafting a unified measure of (among other things) agricultural yields in America in the 1930s around the time of the Great Depression. In his analysis, quantification allowed a grasp of "America as a whole," a qualitative phenomenon that was instrumental for governmental interventions (policies) and the welfare state. Thus, it is not the novelty but the versatility (Kallinikos et al. 2013; Zittrain 2006; Yoo et al. 2010; see chapter 6), resulting in expansion in scope and reach, that characterizes digitalization qua quantification. In what follows I discuss the emergent picture of digitalization through the three dimensions of this book outlined in chapter 1, the objects, modes, and machineries of knowing.

OBJECTS OF KNOWING: PHENOMENOLOGICAL REAL

Understanding digitalization qua quantification implies pursuing Boellstorff's (2016) call to understand how the digital can be real by tracing out how the "real" physical objects that intentionally oriented, purposeful work practices are directed toward are rendered digitally. Part II of this book provides empirical detail—across real physical objects such as oil reservoirs in geological formations below the seabed, sand particles gushing through pipelines of extracted hydrocarbons, and fish, mammals, and corals as part of the Arctic marine environment—of the processes that go into making digital renderings meaningful and consequential. Collectively, the empirical accounts in part II portray how IoT data feeds increasingly conjure up the phenomenological lifeworld in organizationally meaningful ways; that the digital representations and algorithmic phenomenon are organizationally real. This is in part tied to the increasing richness or scope of sensors' ability (or rather, attempts) to quantify qualitative, tactile experiences, including smell, taste, temperature, and visual perception (Singh et al. 2014). In addition, the real-time and interactivity characteristics add to the realism of the digital, as suggested by the notion of nowcasting (Constantiou and Kallinikos 2015). As Knorr Cetina (2009) notes, the real-time tickers of traders in her case create a fluid reality, a synthetic situation in which the digital representations (ticker feeds) are perceived as real because "as the [real-time] information scrolls down the screens and is replaced by new information, a new market situation—a new reality—continually projects itself" (72).

Again, these processes of quantifying qualitative objects need to be understood as efforts or attempts at quantification, not accomplishments. A helpful way to unpack the tensions and trade-offs implicated in these efforts is through Latour's (1999) notion of a *circulating reference*. The notion is illustrated by an ethnographic study of a team of life scientists targeting the qualitative, physical object of a particular section of the Amazon forest in Boa Vista, Brazil. They are tasked with determining whether the savanna is retreating or expanding in this area of the Amazon forest, a question no different in character from that in chapter 7 regarding potential disturbances to the Arctic marine environment. In Latour's account, the scientists go about answering the question through a sequence of translations quantifying aspects of the originating qualitative phenomenon. The undifferentiated forest is partitioned into a grid of equal-sized squares, which is one-to-one mapped and miniaturized to the scale of a small box divided into the same grid pattern. Physical samples from the gridded forest fill corresponding cells in the miniature, known as a *pedocomparator*. The samples are subsequently quantified by measures of color (using the Mansell code) and composition. There are, in the context of this book, two salient aspects of Latour's analysis.

First, the successive steps of quantification outlined above are not about replacing or substituting the qualitative for quantified renderings, but translating. The forest is not the same as the miniature box mapping it, but for the purpose of answering their question about whether the savanna is retreating, it may fill a productive role. This is similar, for instance, to Prentice (2013), who studied surgeons. For the particular purpose of teaching surgical procedures, she found that the embodied, tactile knowledge work of surgeons could be replaced by digitally rendered substitutes in which "[real-time] graphics [of surgery] replace the sense of 'hands-on'" (83). This is exactly the point emphasized in this book when, for instance, digital renderings of sand (sensor measurements, plotted trends, predictive algorithms; see chapter 6), for the purpose of sand-monitoring routines, fill the role of physical sand. Tying back to Boellstorff's (2016) call for understanding the realness of the digital, then, amounts to analyzing the conditions and purposes, be it a retreating versus expanding Amazon forest, the search for oil reservoirs in geological formations, or the study of the biomass in the Arctic, for practically useful translations of quantified/ digital representations. It is, in other words, an epistemic concern for conditions for performing tasks rather than an ontological concern for what these representations "really" are.

Second, and building on that above, the translations/steps of quantifications are motivated by pragmatic concerns or trade-offs: the translations are symmetric in the sense that you gain something but simultaneously lose something. Relative to purpose, the translations are helpful and therefore attractive to engage with, or not. In Latour's case, the qualitative richness of the originating phenomenon (Amazon forest) is traded for increased mobility. Similarly, in the empirical cases in this book the qualitative richness of geological formations, sand, or the marine environment is traded for quantified digital renderings for practical, goal-directed organizational purposes.

MODES OF KNOWING: SCAFFOLDING

Qualitative judgment, interpretation, and sensemaking are constitutive aspects of human reasoning or "intelligence" (Dreyfus and Dreyfus 2000). These characteristics are central to arguments about the so-called knowledge-intensive work practices related to the delegation of tasks to technology (Autor 2015). Given an understanding of digitalization in general and data-driven approaches in particular as efforts of quantifying the qualitative, the issue of limits to automation and *human-in-the-loop* takes center stage (Mindell 2015; Shrestha et al. 2019). Rather than offering an abstractly defined, closed formula of the boundary of qualitative/quantitative, what this book provides is an account of the collective, hybrid achievement of uptake into organizational action and decision-making. As von Krogh (2018) notes, "How problem-solving with the involvement of intelligent machines unfolds in organizations remains a poorly understood phenomenon" (406). With increasingly imperialistic tendencies, "data science is portraying itself as a *universal science*" (Ribes 2019, 517). Thus, data-driven efforts need to be conceptualized as fallible projects that may or may not work out for specific purposes and situations; they are performed achievements (Callon 2007; Pickering 2010; MacKenzie 2006). The emergent picture is one in which, on the one hand, datafication for a given purpose may work in practice but not necessarily in theory (to paraphrase LaPorte and Consolini [1991]), while, on the other hand, apparent "automation" (hence quantification) amounts to relocating and/or transforming, not eliminating, the role of the qualitative (Bechmann and Bowker 2019). Berg and Timmermans (2000) make a related argument when arguing that "these orders do not emerge out of (and thereby replace) a pre-existing disorder. Rather, with the production of an order, a corresponding disorder comes into being" (36–37). Hence, automation, expressed informally, is like the game of whack-a-mole: every time you eliminate it in one place, it keeps reappearing elsewhere. An illustrative example is the formation and curation of training sets for data-driven methods.[1] Computer vision relies heavily on supervised algorithms for most applications (Bechmann and Bowker 2019). Generating training sets for supervised algorithms requires significant data work in the form of expert-based, manual labeling. As a consequence, training sets are in high demand and tend to draw extensively on widely available benchmark training sets. In the case of visual perception, this would typically involve the almost ten-year-old and relatively small ImageNet data set (Deng et al. 2009). Yet, "the datasets which machine learning (ML) critically depends on—and which frequently contribute to errors—are often poorly documented, poorly maintained, lacking in answerability, and have opaque creation processes" (Hutchinson et al. 2021, 560).

Similarly, within an information-processing, microeconomic perspective, decision-making can be separated into predictions, which in such a perspective is the production of information you do not have from what you do have, and judgments, which weigh or assess the value of identified predictions. Viewed through this lens, the former—but, crucially, not the latter—is amendable to quantification through data science (Agrawal et al. 2018; see also Shrestha et al. 2019).

A helpful way to think about the emergent hybrid, collective arrangement of data-driven approaches to organizational decision-making is through the notion of *scaffolding* as developed by Wylie from studies in archaeology (see Wylie 2017; Wylie and Chapman 2014; Chapman and Wylie 2014). It offers a way to theoretically characterize the practically working quantification of qualitative sensemaking involved in digital oil. The domain of archaeology shares a number of similarities with oil. Of the different phases of commercial oil discussed in this book, the phase of exploration covered in chapters 4 and 5 is particularly close to Wylie's account of archaeology: knowledge is partial, provisional, fallible, and influenced by the arrival of quantified measurement techniques (including carbon-14 isotope decay, lead isotope analysis, dental enamel for oxygen isotopes). The scaffolding of archaeological knowing builds, and continuously rebuilds, credible background knowledge to develop and mobilize meaningful interpretations of the material evidence, juggling several interpretations (or working hypotheses) at the same time.

Consistent with a performative and relational perspective, "archaeological facts," exactly like facts in digital oil, grapple with the problem "that the tangible, surviving facts of the record so radically underdetermine any interesting claims archaeologists might want to make that archaeologically based 'facts of the past' are inescapably entangled with fictional narratives of contemporary sense-making" (Wylie 2010, 301). Hence, in the case of oil exploration quantified, real-time, IoT data are only meaningful against a backdrop of a qualitative, narrative understanding of geological processes or history.

Furthermore, scaffolding is provisional. In archaeology, as in the geosciences, there is significant competence in moving hermeneutically between close-up, measured data points and taking a step back to gain an appreciation of the broader formative processes: "[Archaeologists] have built up a repertoire of research strategies specifically designed to mobilise the evidence of human lives and events that survives in an enormous range of material evidence . . . putting material evidence to work in the investigation of a great many different aspects of the cultural past" (Chapman and Wylie 2014, 5). In the case of digital oil, the many ways that data are corroborated, triangulated, and calibrated, as described in part II, are similarly the principal ways of moving back and forth between the micro and the macro.

Scaffolding, Chapman and Wylie (2014) point out, is decentered, distributed, and collective. Scaffolding involves "technical expertise and community norms of practices which are internalized by individual practitioners as embodied skills and tacit knowledge, and externalized in the material and institutional conditions that make possible the exercise, and the transmission of these skill and this knowledge" (55). Scaffolding, in other words, needs to be understood through the infrastructure lens of the machineries of knowing, to which this book adheres. Taken together, scaffolding—dynamic, provisional, decentered—frames the performed achievement of the organizational knowing of digital oil, oscillating between quantified and qualitative expressions.

MACHINERIES OF KNOWING: "BRINGING WORK BACK IN"

The datafication of society—the digitalization of "everything"—has been observed by several scholars (Lycett 2013; Markus 2017; Leonelli 2014; Kallinikos et al. 2013). Its infrastructure leans on the *platformization* of services and offerings promoting scope and scale (Cusumano et al. 2020; Gawer 2011). The literature on digital platforms originates from new innovation or business operation models, as exemplified by Apple's iOS (Eaton et al. 2015), Google's Search ecosystem (Iyer and Davenport 2008), and Facebook (Rogers 2016). Capitalizing on the way digital platforms enjoy economy of scale alongside a capacity to specialize has led to commercial success for a variety of devices and services, including smartphones (Eaton et al. 2011), advertisements (Alaimo and Kallinikos 2018), social media (Plantin et al. 2018), and wearable technologies (Schüll 2016).

Social media platforms are prominent expressions of how the network externalities of an increasing user base and services drive the evolution of platforms (Ford and Wajcman 2017; Plantin et al. 2018; Stark 2018; Alaimo and Kallinikos 2018). Users' behavior, attitudes, and preferences and an expanding list of other qualitative characteristics about us are quantified through data traces and algorithmic constructs. Gerlitz and Helmond's (2013) study of the "like economy" is illustrative. It unpacks the formation of and the machineries behind algorithmic constructs (such as reputation, connectivity) that nudge or manipulate the online behavior that feeds the organization and structure of social media platforms. A central illustration is the Like button on Facebook:

> The button provides a one-click shortcut to express a variety of affective responses such as excitement, agreement, compassion, understanding, but also ironic and parodist liking. . . . By asking users to express various affective reactions to web content in the form of a click on a Like button, these intensities can be transformed into a number on the Like counter and are made comparable. (Gerlitz and Helmond 2013, 1358)

Zuboff's (2019) analysis of surveillance capitalism provides a compelling understanding of how platforms feed off, not to mention exploit, our *behavior surplus*, the digital traces of our behavior that we, the users, are seduced to give up in exchange for attractive services.

For all its merit, there is in dominant accounts of the platformization of datafication a strong yet largely implicit and unchallenged assumption of *individualized consumer choice* (exceptions include Bonina et al. 2021). Dominant ways of conceptualizing platformization, epitomized through social media platforms, are culturally and materially shaped by its inception as vehicles to serve mass consumption markets. Thus, the quantification of qualitative behavioral surplus, to use Zuboff's phrase, assumes—and is limited by—a form of methodological individualism. In neoliberal, consumer capitalist societies, "[Individual] choice is a sine qua non of contemporary life. . . . Platforms are not simply cameras that present choice and enable comparisons between different options, but are more akin to engines that govern, drive and expand choice, configuring users within particular discourses, practices and subjectivities" (Graham 2018, 1). Consumer choice is not merely about consumption but, crucially, cultural expression of self-identity—that is, *Homo eligens* (Bauman 2007). As Kotliar (2020) notes, "Our choices are fast becoming algorithmic. The ubiquity of recommender systems, personalization engines, and user analytics services has made algorithms almost inseparable from our everyday choice-making" (347).

In the context of this book, however, this dominant emphasis on the conceptualization of platformization on consumerism is problematic. It leaves out what is fundamental to this book—namely, the social organizing and institutional fabric in which this platformization/infrastructuring unfolds. In short, the *social* aspects of work—shaped by datafication and fueled by platformization—are left unaccounted for, with an individualistic and atomistic quantification of users. For grasping the role of platformization in promoting and expanding datafication, "bringing work back in" is necessary (Barley and Kunda 2001). The proclaimed, radical transformational capacity of digital platforms—the "Uberization" of organizations (Faraj and Pachidi 2021)—relies heavily on the quantification of atomistic, individualized users,[2] not to mention consumers (Kotliar 2020). In contrast, the platformization of business-to-business or organizational exchange, where users are not atomistic but belong to organizational collectives with dependencies between users' tasks, has had a much slower uptake. Part II of this book vividly demonstrates how organizational and institutional aspects shape the uptake of platformization qua quantification in ways that the prevailing literature, emphasizing individualistic or consumer choice, fails to capture. In other words, the impact of AI and data science is strongest when coupled with platforms and when platforms capture (quantify) atomistic user behavior, with organizational uptake of those same technologies markedly slower (Günther et al. 2017).

Furthermore, if previously the focus was on the theory of the firm (Cyert and March 1963), the significance of comprehensive ecosystems shifts the attention to theorizing about industry-wide transformation (Geels and Schot 2007); it shifts the *unit of analysis* for studying digitalization. Digitally enabled changes during the last many years have focused on intraorganizational change (Vial 2019). The changes heralded by digital platforms pertain to complete industrial ecosystems. This book illustrates this point. The changes analyzed result from broad, mutually reinforcing initiatives to employ digitalization within the whole ecosystem of oil operators, oil service providers, the engineering, procurement, and construction companies, and vendors and consultants of digital solutions, as well as public authorities and agencies. Scholars of digitalization, then, are well advised to focus on the transformation of whole industries, breaking away from traditional case studies of singular organizations (Williams and Pollock 2012).

In conclusion, reading digitalization through the lens of the attempted quantification of quality, as proposed by this book, is consistent with a commitment to empirically open, analytically critical phenomenon-oriented theorizing (von Krogh 2018). Rather than an ideological, philosophical, or otherwise given boundary between the qualitative and the quantitative, a dynamically evolving, uneven, fallible, and varied landscape opens up, a landscape offering rich opportunities for further scholarly travels.

Appendix: A Note on Method

The empirical material in this book is the result of my more than two decades of engagement with the industrial ecosystem around Norwegian offshore oil and gas. Across a number of research projects, with a varied set of collaborating researchers, PhD students, and postdocs, I have studied aspects of digitalization in the industry. The details of data collection and data analysis in these projects are provided in the papers published with my coauthors. Here, I offer some reflections of a methodological nature beyond what is addressed in those papers.

First, the longitudinal research design in this book is not something I planned but rather a gradually acquired quality. I did not aim for a biography of artifact (Williams and Pollock 2012) or a longue durée (Ribes and Finholt 2009) perspective. In the political economy of my academic environment, there is a clear expectation that we generate external research funding for PhDs and postdocs. For idiosyncratic reasons, some of the early research funding I secured happened to be within the domain of oil and gas. Securing additional research funding is always easier than starting from scratch. I thus somewhat opportunistically stumbled into the digitalization of oil and gas. However, remaining and digging further into it certainly was not by chance. I was motivated by a clear sense that there was *something* empirically here, which I struggled to wrap my head around in a phenomenon-driven mode of theorizing (von Krogh 2018).

Second, my engagement with oil does indeed span an extended period, but with huge variations in intensity. Over the years I have in parallel had an equally extended empirical engagement with digitalization in the health-care sector. Across several research projects, I have, together with collaborators, explored digitalization in secondary care in hospitals, in primary care in municipalities, in pharmacies, and with general practitioners. The fieldwork is dominated by the Norwegian context but also includes studies of health care in Africa and India. It is difficult to come up with two empirical domains more different than public health care and corporate oil and gas. Personally, moving between them has been a source of inspiration, energy, and necessary variation. Despite the differences at the empirical level, however, I have pursued a form of triangulation at the level of conceptualizing and theorizing. Hence, the broad characterization of digitalization promoted in this book resonates deeply with tendencies also found in in health care: liquefaction, or disembedding, is manifest in the sensor- and Internet of Things–based push for so-called welfare technologies (Grisot et al. 2019);[1] interest in data-driven approaches tap into efforts toward more evidence-based medicine (Timmermans and Berg 2010); and, increasingly, digital platforms are recognized as key to orchestrating an evolving set of health services (Hanseth and Bygstad 2015).

Third, my personal preference was always for relatively small, light-weight research projects devised and conducted directly with implicated stakeholders in partner organizations. The unceremonious manner in which these projects can typically be set up locally with partners in public and private organizations has been a comparative advantage enjoyed by many Scandinavian researchers. This is changing, if not in general, certainly within the oil and gas industry. Research projects need to go through increasingly bureaucratic procurement processes requiring blessings from upstairs, which has consequences for the nature and size of the projects. Against my natural instincts, I thus find myself increasingly working in large research consortiums or networks with a wide range of researchers, only a fraction of whom would be familiar with my own research tradition. For instance, I am currently involved in two large research projects in the oil domain, each with twenty-plus researchers and still more PhDs/postdocs: Sirius (2021) on big-data access in oil and BRU21 (2021) on digitalization given a scenario of oil prices below thirty US dollars per barrel. Large, nonorganic research efforts are not really my cup of tea. I need there to be an organic element based on professional trust and respect, either present from the start (as with Sirius, led by someone who was a graduate student together with me ages ago) or cultivated over time (as with BRU21).

In sum, the mode of research underpinning this book evolved from fairly autonomous, self-sufficient, and modestly sized to larger consortia-based engagements. This shift mirrors in part changes in how research funding is organized in the domain under study, corporate oil and gas. In addition, however, it reflects more generally how studying increasingly "large" phenomena—climate change monitoring (Edwards 2010), long-term biological diversity (Karasti et al. 2006), or, the theme of this book, the expansive if not imperialistic nature of digitalization—increasingly leans on more collective and long-term ways of organizing research.

Notes

CHAPTER 1

1. The vision attracts interest but, more importantly, investments. The world's first life-size subsea gas compressor prototype has successfully been completed (see Nilsen 2015).

2. Clearly, this is a simplification. Form has content. As Dourish (2017) illustrates with different representations of numbers, Indo-Arabic representations of numbers lend themselves to arithmetic procedures for addition, subtraction, multiplication, and division in ways Roman representations of numbers do not. Hence, especially from the perspective of knowledge infrastructures, this demonstrates the epistemic differences tied to differences in representations of numbers. Despite the simplification of the reference/referent dichotomy, it is useful for outlining key concerns with digitalization.

3. This resonates with a comment Susan Leigh Star once made at a seminar but never (as far as I know) put into writing that the relationship between organization and IT is the same as that between institution and infrastructure. Or as Rob Kling similarly once commented (again, without ever writing about it), design relates to individual houses as infrastructures do to urban planning.

4. Using the adjective (*algorithmic*) rather than the noun (*algorithm*) is consistent with arguments in critical AI of avoiding the essentialist, artifact-centric conceptualization of data science (Bucher 2018; Gillespie 2016; Glaser et al. 2021).

5. See, e.g., Inductive Automation 2018.

6. Chapter 4 was written in collaboration with Marius Mikalsen. Some parts of it have been published in Mikalsen and Monteiro (2018).

7. This is a cartoon version. For instance, field development, which consists of devising the production method; planning the production facilities; and, not least, determining how a new field will tap into the existing infrastructure of pipelines, processing capacity, and refineries, is hardly straightforward.

8. Chapter 5 was written in collaboration with Marius Mikalsen. Parts of it, in an earlier version, have been published as Mikalsen and Monteiro (2021).

9. Chapter 6 was written in collaboration with Thomas Østerlie. Parts of it, in an earlier version, have been published as Østerlie and Monteiro (2020).

10. There is an ongoing debate on "what Norway should live off of after oil." See, e.g., Støa 2020.

11. Chapter 7 was written in collaboration with Elena Parmiggiani. Parts of it, in an earlier version, have been published as Monteiro and Parmiggiani (2019). In addition, a few details have in earlier versions been published in Parmiggiani et al. (2015).

CHAPTER 2

1. For instance, the recent oil discovery in the Norwegian Sea has been planned with a minimal time from discovery to production (NRK 2021).

CHAPTER 3

1. A senior exploration scientist with a major oil operator once confessed that he was increasingly concerned about whether new generations of geologists had sufficient appreciation and knowledge of the physical phenomena, given that field trips to analogues are dwindling in frequency in commercial companies.

2. Other important activities include field development, facility management, and process engineering.

3. *Explorationist* is used as a collective term for the geologists and geophysicists (and other geoscience disciplines, including geochemists). They refer to themselves as *interpretationists* or simply as *G&G*, short for "geology and geophysics." This naming signals the multidisciplinary effort of crafting geological interpretations of the subsurface.

4. Several technologies from different vendors are exploring this, one of which is IntelliServ.

5. The slogan-like characterization into a four-V model of big data was proposed by IBM, supplementing an earlier model by Gartner that lacked attention to "veracity," a defining aspect of IoT data; see, e.g., Perry 2017.

6. In recognition of this, several operators have worked out new contracts that stipulate economic sanctions for poor-quality data from the drillers and thus are inscribing incentives for increased data quality into the contracts between the operators and the drilling companies.

7. A few years ago, NorthOil hired one of the large technology providers to help it set up an effective search tool for its intranet based on crawling and indexing its documents and files. As one informant laughingly explained, the search engine could not figure out the correct access regimes. When one of the informant's colleagues tried typing the key word "confidential" into the search engine, "the result was all confidential documents." This resulted in the entire intranet being shut down for three months until the problem was resolved.

8. For instance, as part of the efforts toward integrated operations (see chapter 1), NorthOil formed production optimizing teams consisting of colocated (physically sharing a large desk)

production engineers with short-term (hours and days) time frames and reservoir engineers with longer-term (decades, the life span of an oil field) time frames.

9. There is a time delay of a few years for seismic interpretations.

10. Several exist, including one by Petrovisor.

11. The OSDU initiative was kick-started by Schlumberger and Shell open sourcing key parts of their technology; see OSDU 2021.

12. In Norwegian offshore oil, the NORSOK standard is among the most important; see Standards Norway 2021.

CHAPTER 4

1. This is not to deny the considerable opposition and critique from many of the companies selling these ads, such as when Procter and Gamble, the biggest buyer of commercial ads worldwide, voices its frustration with some of the results of programmatic advertising. See Handley 2017.

2. Jackson uses the notion of articulation work as a form of invisible work.

3. There are exceptions. The Norwegian Petroleum Directorate, for instance, will commission seismic surveys on its own if and when it believes an area needs surveying but has yet to be surveyed by any of the commercial actors.

CHAPTER 5

1. There are, e.g., a number of open-source tools available, including Hadoop, Apache Spark, MapReduce, Cassandra, MongoDB, and Tableau.

2. As Marcus (2018) reminds us, the qualifier "deep" refers to the number of hidden layers in a neural network. It is thus a technical feature of the technology, not to be mistaken for psychological "deep" learning.

3. Similarly, work on explainable AI and XAI (explainable AI) explores methods to enhance the transparency, and hence the accountability, of black-boxed AI methods (Miller 2019).

4. Vitrinite reflectance is a measurement of the optical properties of vitrinite, a form of organic matter contained in rock samples. Vitrinite is used to diagnose the maturity of source rock, as its reflectance is sensitive to temperature ranges that correspond to those of hydrocarbon generation.

CHAPTER 6

1. Leaking gas is considered among the most dangerous situations on an offshore petroleum installation because of the catastrophic consequences of gas explosions (Kongsvik et al. 2011). While the 2010 Deepwater Horizon catastrophe in the Gulf of Mexico was not caused by a punctured pipeline, it illustrates the twin dangers of an uncontrolled oil spill causing

disastrous environmental damage and the consequences of leaking gas that ignites and subsequently explodes the topside platform.

2. The term "digitalization" has many meanings. With automation as the formative idea, studies of digitalization in the 1980s and 1990s were predominately conceptualized as *computerization* (Kling 1996). Gradually acknowledging the expanded depth and width implied in embracing the transformative capacity, there has since been a proliferation of concepts to capture digitalization beyond the automation/substitution tied to the concept of computerization. "Virtual/ization" is one widely used term. Some use it loosely to denote when physical mechanisms or processes are conducted by computers rather than physically (Overby 2008) or where face-to-face communication is mediated by computers (Jarvenpaa et al. 1998). In contrast, Bailey et al. (2012) provide a definition in which they identify digitization (the creation of computer-based representations of physical phenomena) as a necessary precursor to and hence different from virtuality—i.e., the engagement with these representations. This useful clarification corresponds to Yoo et al.'s (2010) distinction between digitization as the coding into digital formats and digitalization as the processes of engagement made possible by digitization. Here, Bailey et al.'s (2012) notion of the virtual is adopted.

3. The taxonomy into three types is found, without resorting to Peirce, in Knorr Cetina (1999). She identifies three types of data. Physical phenomena, first and traditionally, may be staged to produce data that correspond with the phenomena directly (i.e., the indices in Peirce). Second, the physical conditions are manipulated to yield processed, partial versions of data that are equivalent or similar (i.e., icons in Peirce). Third, and most radically, physical phenomena are mere signatures and footprints of events, providing data as signs (i.e., symbols).

4. This and later names are anonymized.

CHAPTER 7

1. Interestingly, the opposition is escalating from within the industry, and not only outside. For instance, in their annual report the International Energy Agency, historically a strong supporter of the fossil fuel industry, issued a remarkably strong conclusion that the "exploitation and development of new oil and gas fields must stop this year" (see Harvey 2021) and ExxonMobil, against the chairpersons, has had to accept the arrival of three new board members widely viewed as proponents of an environmentally friendly agenda.

2. Clearly, this is a simplification and not quite as naively technocratic as it may seem. Yet it acts as a navigating ideal that significantly shapes the procedure and content of the political processes.

3. The figure includes oil reserves only, not natural gas. For many of the fields, there are more gas than oil reserves. The main point, however, still applies—namely, that there is in the oil industry a mounting concern about the lack of untapped hydrocarbon reserves, which adds to their appetite for new areas in the Arctic.

4. An interactive map of environmental "values" jointly created by the Norwegian Directorate of the Environment and the Mapping Agency (Kartverket) inspired the notion in EnviroTime of the value of biomass; see Barentswatch 2021.

5. The lobbying succeeded. In the twenty-third concessional round announced in 2016, ten new licenses were opened for oil and gas activities, three of these in the Norwegian part of the Barents Sea. A coalition of environmental NGOs sued the government, claiming that the twenty-third round went against the rights to a healthy environment written into the Constitution. The case was lost in the first level of the courts in January 2018. Oil exploration in Venus is still banned.

CHAPTER 8

1. More formally, this insight may be formulated by underscoring the "slight surprise of action" (Latour 1999) or, alternatively, the way side effects and unintended outcomes overtake the intended ones (Beck 1992; Perrow 2011).

2. Zuboff (2019) is on to the same idea when pointing out how Skinner's behaviorism, which similarly relies on methodological individualism, is an illuminating way to understand the perspectives on users' behavior (implicitly) pursued by the big-tech platform companies. What Zuboff fails to discuss, however, is the role of platforms in organizational and institutional settings where users are not atomistic.

APPENDIX

1. Welfare technologies is a notion also known as ambient technologies. It consists of a variety of largely sensor-based services for monitoring health conditions, including not only blood sugar levels and respiration but also patient safety services such as fall detectors and GPS tracking of Alzheimer patients.

References

Abrahamson, E. (1991). Managerial fads and fashions: The diffusion and rejection of innovations. *Academy of Management Review, 16*(3), 586–612.

Agrawal, A., Gans, J., & Goldfarb, A. (2018). *Prediction machines: The simple economics of artificial intelligence*. Cambridge, MA: Harvard Business.

Aker BP. (2021). *Ivar Aasen*. Retrieved October 15, 2021, from https://akerbp.com/en/asset/ivar-aasen-3/

Alac, M. (2011). *Handling digital brains: A laboratory study of multimodal semiotic interaction in the age of computers*. Cambridge, MA: MIT Press.

Alaimo, C., & Kallinikos, J. (2018). Objects, metrics and practices: An inquiry into the programmatic advertising ecosystem. In *Working Conference on Information Systems and Organizations* (pp. 110–123). Cham, Switzerland: Springer.

Alaimo, C., & Kallinikos, J. (2020). Managing by data: Algorithmic categories and organizing. *Organization Studies, 41*. https://doi.org/10.1177/0170840620934062

Alavi, M., & Leidner, D. E. (2001). Knowledge management and knowledge management systems: Conceptual foundations and research issues. *MIS Quarterly, 25*(1), 107–136.

Algan, G., & Ulusoy, I. (2020). *Label noise types and their effects on deep learning*. arXiv preprint arXiv:2003.10471

Almklov, P. G., & Hepsø, V. (2011). Between and beyond data: How analogue field experience informs the interpretation of remote data sources in petroleum reservoir geology. *Social Studies of Science, 41*(4), 539–561.

Andersen, H. W. (1997). Producing producers: Shippers, shipyards and the cooperative infrastructure of the Norwegian maritime complex since 1850. In C. Sabel & J. Seitlin (Eds.), *World of possibilities: Flexibility and mass production in Western industrialization* (pp. 461–500). Cambridge, UK: Cambridge University Press.

Anderson, C. (2008). The end of theory: The data deluge makes the scientific method obsolete. *Wired* 16(7). https://www.wired.com/2008/06/pb-theory/

Appadurai, A. (1996). *Modernity at large: Cultural dimensions of globalization*. Minneapolis, MN: University of Minnesota Press.

Arnadottir, A. (2016, May 18). *Regjeringen ofrer iskanten* [The government gives up the ice edge]. Bellona. Retrieved October 15, 2021, from https://bellona.no/nyheter/olje-og-gass/2016-05 -regjeringen-ofrer-iskanten

Asaro, P. M. (2000). Transforming society by transforming technology: The science and politics of participatory design. *Accounting, Management and Information Technologies, 10*(4), 257–290.

Ash, J. S., Berg, M., & Coiera, E. (2004). Some unintended consequences of information technology in health care: The nature of patient care information system-related errors. *Journal of the American Medical Informatics Association, 11*(2), 104–112.

Autor, David. (2015). Why are there still so many jobs? The history and future of workplace automation. *Journal of Economic Perspectives, 29*(3), 3–30.

Bachelard, G. ([1949] 1998). *La Rationalism Appliqué*. Paris: Presses Universitaires de France.

Bailey, D. E., Leonardi, P. M., & Barley, S. R. (2012). The lure of the virtual. *Organization Science, 23*(5), 1485–1504.

Baldwin, C. Y., & Clark, K. B. (2003). Managing in an age of modularity. In R. Garud, A. Kumaraswamy, & R. N. Langlois (Eds.), *Managing in the modular age: Architectures, networks, and organizations* (pp. 149–160). Oxford, UK: Blackwell.

Barad, K. (2003). Posthumanist performativity: Toward an understanding of how matter comes to matter. *Signs, 28*(3), 801–831.

Barentswatch. (2021). *Arealverktøy for forvaltningsplanene* [Arial tool for policy-making]. Retrieved October 15, 2021, from https://kart.barentswatch.no/

Barley, S. R. (1986). Technology as an occasion for structuring: Evidence from observations of CT scanners and the social order of radiology departments. *Administrative Science Quarterly, 31*, 78–108.

Barley, S. R., & Kunda, G. (2001). Bringing work back in. *Organization Science, 12*(1), 76–95.

Barrett, M., Davidson, E., Prabhu, J., & Vargo, S. L. (2015). Service innovation in the digital age: Key contributions and future directions. *MIS Quarterly, 39*(1), 135–154.

Baskerville, R. L., Myers, M. D., & Yoo, Y. (2020). Digital first: The ontological reversal and new challenges for IS research. *MIS Quarterly, 44*(2), 509–523.

Bateson, G. (1972). *An ecology of mind*. New York, NY: Ballantine.

Baudrillard, J. (1994). *Simulacra and simulation*. Ann Arbor, MI: University of Michigan Press.

Bauman, Z. (2007). *Consuming life*. Cambridge, UK: Polity Press.

Bechmann, A., & Bowker, G. C. (2019). Unsupervised by any other name: Hidden layers of knowledge production in artificial intelligence on social media. *Big Data & Society, 6*(1). https:// doi.org/10.1177/2053951718819569

Beck, U. (1992). *Risk society: Towards a new modernity.* London, UK: Sage.

Berg, M., & Timmermans, S. (2000). Orders and their others: On the constitution of universalities in medical work. *Configurations, 8*(1), 31–61.

Beunza, D., & Garud, R. (2007). Calculators, lemmings or frame-makers? The intermediary role of securities analysts. *Sociological Review, 55*(2), 13–39.

Biello, D. (2015, April 28). How microbes helped clean BP's oil spill. *Scientific American, 28.* https://www.scientificamerican.com/article/how-microbes-helped-clean-bp-s-oil-spill/

Bird, K. J., Charpentier, R. R., Gautier, D. L., Houseknecht, D. W., Klett, T. R., Pitman, J. K., Moore, T. E., Schenk, C. J., Tennyson, M. E., & Wandrey, C. J. (2008). *Circum-Arctic resource appraisal: Estimates of undiscovered oil and gas north of the Arctic circle.* Fact sheet no. 2008–3049. US Department of the Interior, US Geological Survey. http://pubs.usgs.gov/fs/2008/3049/

Bjørnestad, S. (2019, November 4). *Offentlig utvalg foreslår særskilt skatt på oppdrett* [A public group of advisers suggests targeting fish farming with taxes]. Aftenposten. Retrieved October 15, 2021, from https://www.aftenposten.no/okonomi/i/OpXlJb/offentlig-utvalg-foreslaar-saerskilt -skatt-paa-oppdrett

Blanchard, A., Hauge, K. H., Andersen, G., Fosså, J. H., Grøsvik, B. E., Handegard, N. O., Kaiser, M., Meier, S., Olsen, E., & Vikebø, F. (2014). Harmful routines? Uncertainty in science and conflicting views on routine petroleum operations in Norway. *Marine Policy, 43*(January), 313–320.

Bloom, J. (2019, March 8). *Norway's $1 trillion fund to cut oil and gas investments.* BBC. Retrieved October 15, 2021, from https://www.bbc.com/news/business-47494239

Bobadilla, J., Ortega, F., Hernando, A., & Gutiérrez, A. (2013). Recommender systems survey. *Knowledge-Based Systems, 46,* 109–132.

Boellstorff, T. (2016). For whom the ontology turns: Theorizing the digital real. *Current Anthropology, 57*(4), 387–407.

Bojarski, M., Del Testa, D., Dworakowski, D., Firner, B., Flepp, B., Goyal, P., Jackel, L., et al. (2016). *End to end learning for self-driving cars.* arXiv preprint arXiv:1604.07316

Bond, C. E. (2015). Uncertainty in structural interpretation: Lessons to be learnt. *Journal of Structural Geology, 74,* 185–200.

Bonina, C., Koskinen, K., Eaton, B., & Gawer, A. (2021). Digital platforms for development: Foundations and research agenda. *Information Systems Journal.* Forthcoming.

Borgman, C. L., Edwards, P. N., Jackson, S. J., Chalmers, M. K., Bowker, G. C., Ribes, D., Burton, M., & Calvert, S. (2013). Knowledge infrastructures: Intellectual frameworks and research challenges. https://escholarship.org/uc/item/2mt6j2mh

Borgmann, A. (1999). *Holding on to reality: The nature of information at the turn of the millennium.* Chicago, IL: University of Chicago Press.

Bowker, G. C. (1994). *Science on the run: Information management and industrial geophysics at Schlumberger, 1920–1940.* Cambridge, MA: MIT Press.

Bowker, G. C. (2014). Big data, big questions: The theory/data thing. *International Journal of Communication, 8*, 1795–1799.

Bowker, G. C., & Star, S. L. (2000). *Sorting things out: Classification and its consequences.* Cambridge, MA: MIT Press.

Braverman, H. (1974). *Labor and monopoly capital: The degradation of work in the twentieth century.* New York, NY: New York University Press.

Brown, D. (2021, March 23). Why it will be years before robot butlers take over your household chores. *The Washington Post.* Retrieved October 15, 2021, from https://www.washingtonpost.com /technology/2021/03/23/future-robots-home-jetsons/

Brown, M. (2021, May 18). "Some people feel threatened": Face to face with Ai-Da the robot artist. *The Guardian.* Retrieved August 1, 2021, from https://www.theguardian.com/culture/2021 /may/18/some-people-feel-threatened-face-to-face-with-ai-da-the-robot-artist

Brown, W. M. (1983). The economy of Peirce's abduction. *Transactions of the Charles S. Peirce Society, 19*(4), 397–411.

BRU21. (2021). *BRU21: Research and innovation program in digital and automation solutions for the oil and gas industry.* Retrieved August 1, 2021, from https://www.ntnu.edu/bru21

Brynjolfsson, E., & Hitt, L. M. (2000). Beyond computation: Information technology, organizational transformation and business performance. *Journal of Economic Perspectives, 14*(4), 23–48.

Brynjolfsson, E., & McAfee, A. (2014). *The second machine age.* New York, NY: W. W. Norton.

Bucciarelli, L. (2003). *Engineering philosophy.* Delft, Netherlands: DUP Satellite; an imprint of Delft University Press.

Bucher, T. (2018). *If . . . then: Algorithmic power and politics.* Oxford, UK: Oxford University Press.

Burton-Jones, A. (2014). What have we learned from the smart machine? *Information and Organization, 24*(2), 71–105.

Burton-Jones, A., & Grange, C. (2013). From use to effective use: A representation theory perspective. *Information Systems Research, 24*(3), 632–658.

Busch, L. (2011). *Standards: Recipes for reality.* Cambridge, MA: MIT Press.

Callon, M. (2007). What does it mean to say that economics is performative? In D. MacKenzie, F. Muiesa, and L. Siu (Eds.), *Do economists make markets? On the performativity of economics* (pp. 311–357). Princeton, NJ: Princeton University Press.

Callon, M., Lascoumes, P., & Barthe, Y. (2011). *Acting in an uncertain world: An essay on technical democracy.* Cambridge, MA: MIT Press.

Carlile, P. R. (2004). Transferring, translating, and transforming: An integrative framework for managing knowledge across boundaries. *Organization Science, 15*(5), 555–568.

Carstens, H. (2014, July 3). *Et mye omtalt brev* [A letter much talked about]. Geo365. Retrieved October 15, 2021, from https://geo365.no/olje-og-gass/et-mye-omtalt-brev/

Cecez-Kecmanovic, D., Galliers, R. D., Henfridsson, O., Newell, S., & Vidgen, R. (2014). The sociomateriality of information systems. *MIS Quarterly, 38*(3), 809–830.

Chapman, R., & Wylie, A. (Eds.). (2014). *Material evidence: Learning from archaeological practice.* London: Routledge.

Chang, H. (2004). *Inventing temperature: Measurement and scientific progress.* Oxford, UK: Oxford University Press.

Chen, C., Seff, A., Kornhauser, A., & Xiao, J. (2015). Deepdriving: Learning affordance for direct perception in autonomous driving. In *Proceedings of the IEEE International Conference on Computer Vision* (pp. 2722–2730). Los Alamitos, CA: Institute of Electrical and Electronics Engineers Computer Society.

Ciborra, C., & Hanseth, O. (1998). From tool to Gestell. *Information Technology & People, 11*(4), 305–327.

Cipolla, C., Gupta, K., Rubin, D. A., & Willey, A. (Eds.). (2017). *Queer feminist science studies: A reader.* Seattle, WA: University of Washington Press.

Constantiou, I. D., & Kallinikos, J. (2015). New games, new rules: Big data and the changing context of strategy. *Journal of Information Technology, 30*(1), 44–57.

Conway, E. M., & Oreskes, N. (2012). *Merchants of doubt.* London, UK: Bloomsbury.

Crosby, A. W. (1997). *The measure of reality: Quantification in Western Europe, 1250–1600.* Cambridge, UK: Cambridge University Press.

Cumbers, A. (2012). North Sea oil, the state and divergent development in the UK and Norway. In J. A. McNeish & O. Logan (Eds.), *Flammable societies: Studies on the socio-economics of oil and gas* (pp. 221–242). London, UK: Pluto Press.

Cusumano, M. A., Yoffie, D. B., & Gawer, A. (2020). The future of platforms. *MIT Sloan Management Review, 61*(3), 46–54.

Cyert, R. M., & March, J. G. (1963). *A behavioral theory of the firm.* Englewood Cliffs, NJ: Prentice Hall.

Davenport, T. (2014). *Big data at work: Dispelling the myths, uncovering the opportunities.* Cambridge, MA: Harvard Business.

de Jonge, B., Teunter, R., & Tinga, T. (2017). The influence of practical factors on the benefits of condition-based maintenance over time-based maintenance. *Reliability Engineering & System Safety, 158*, 21–30.

Deming, D. (2020, January 30). The robots are coming. Prepare for trouble. *The New York Times.* Retrieved October 15, 2021, from https://www.nytimes.com/2020/01/30/business/artificial-intelligence-robots-retail.html

Deng, J., Dong, W., Socher, R., Li, L. J., Li, K., & Fei-Fei, L. (2009). Imagenet: A large-scale hierarchical image database. In *IEEE Conference on Computer Vision and Pattern Recognition* (pp. 248–255). Los Alamitos, CA: Institute of Electrical and Electronics Engineers Computer Society.

DeSanctis, G., & Poole, M. S. (1994). Capturing the complexity in advanced technology use: Adaptive structuration theory. *Organization Science, 5*(2), 121–147.

Dewey, J. (1930). *The quest for certainty.* London, UK: Allen & Unwin.

Didier, E. (2020). *America by the numbers: Quantification, democracy, and the birth of national statistics.* Cambridge, MA: MIT Press.

Dodgson, M., Gann, D. M., & Phillips, N. (2013). Organizational learning and the technology of foolishness: The case of virtual worlds at IBM. *Organization Science, 24*(5), 1358–1376.

Dodgson, M., Gann, D. M., & Salter, A. (2007). "In case of fire, please use the elevator": Simulation technology and organization in fire engineering. *Organization Science, 18*(5), 849–864.

Douglas, M., & Wildavsky, A. (1983). *Risk and culture: An essay on the selection of technological and environmental dangers.* Oakland, CA: University of California Press.

Dourish, P. (2017). *The stuff of bits: An essay on the materialities of information.* Cambridge, MA: MIT Press.

Dreyfus, H., & Dreyfus, S. E. (2000). *Mind over machine.* New York, NY: Simon and Schuster.

Dunne, D. D., & Dougherty, D. (2016). Abductive reasoning: How innovators navigate in the labyrinth of complex product innovation. *Organization Studies, 37*(2), 131–159.

Duportail, J. (2017, September 26). I asked Tinder for my data. It sent me 800 pages of my deepest, darkest secrets. *The Guardian.* Retrieved October 15, 2021, from https://www.theguardian.com/technology/2017/sep/26/tinder-personal-data-dating-app-messages-hacked-sold

E24. (2008, August 18). *Lula gjør som i Norge* [Lula does like Norway]. Retrieved October 15, 2021, from https://e24.no/norsk-oekonomi/i/5V7421/lula-gjoer-som-norge

Eaton, B., Elaluf-Calderwood, S., Sørensen, C., & Yoo, Y. (2011). Dynamic structures of control and generativity in digital ecosystem service innovation: The cases of the Apple and Google mobile app stores. *London School of Economics and Political Science, 44*(0), 1–25.

Eaton, B., Elaluf-Calderwood, S., Sorensen, C., & Yoo, Y. (2015). Distributed tuning of boundary resources: The case of Apple's iOS service system. *MIS Quarterly, 39*(1), 217–243.

Edwards, P. N. (2010). *A vast machine: Computer models, climate data, and the politics of global warming.* Cambridge, MA: MIT Press.

Edwards, P. N., Mayernik, M. S., Batcheller, A. L., Bowker, G. C., & Borgman, C. L. (2011). Science friction: Data, metadata, and collaboration. *Social Studies of Science, 41*(5), 667–690.

Ellingsen, G., & Monteiro, E. (2003). Mechanisms for producing a working knowledge: Enacting, orchestrating and organizing. *Information and Organization, 13*(3), 203–229.

Epstein, S. (1996). *Impure science: AIDS, activism, and the politics of knowledge.* Oakland, CA: University of California Press.

Espeland, W., & Stevens, M. (1998). Commensuration as a social process. *Annual Review of Sociology, 24*(1), 313–343.

Esteva, A., Kuprel, B., Novoa, R., Ko, J., Swetter, S., Blau, H., & Thrun, S. (2017). Dermatologist-level classification of skin cancer with deep neural networks. *Nature, 542*(7639), 115–118.

Faraj, S., & Pachidi, S. (2021). Beyond Uberization: The co-constitution of technology and organizing. *Organization Theory, 2*(1), 2631787721995205.

Faulkner, P., Feduzi, A., & Runde, J. (2017). Unknowns, black swans and the risk/uncertainty distinction. *Cambridge Journal of Economics, 41*(5), 1279–1302.

Feenberg, A. (2012). *Questioning technology.* London, UK: Routledge, 2012.

Feldman, M. S., & March, J. G. (1981). Information in organizations as signal and symbol. *Administrative Science Quarterly, 26,* 171–186.

Fenstad, A., & Hagen, J. M. (2017, August 25). *Over halvparten av sjømatfolket sier olje-nei* [More than half of people involved in the seafood industry are against oil]. Fiskeribladet. Retrieved November 1, 2017, from https://www.fiskeribladet.no/nyheter/over-halvparten-av-sjomatfolket -sier-olje-nei/8-1-54912

Fine, G. A. (2007). *Authors of the storm: Meteorologists and the culture of prediction.* Chicago, IL: University of Chicago Press.

Fiske, A., Prainsack, B., & Buyx, A. (2019). Data work: Meaning-making in the era of data-rich medicine. *Journal of Medical Internet Research, 21*(7), e11672.

Folkeaksjonen. (2017, June 27). *Seismikk—også et problem for hvalen* [Seismic—also a problem for the whale]. Retrieved October 15, 2021, from https://folkeaksjonen.no/content/seismikk-ogsa -et-problem-hvalen

Ford, H., & Wajcman, J. (2017). "Anyone can edit," not everyone does: Wikipedia's infrastructure and the gender gap. *Social Studies of Science, 47*(4), 511–527.

Fosså, J. H., Mortensen, P. B., & Furevik, D. M. (2002). The deep-water coral *Lophelia pertusa* in Norwegian waters: Distribution and fishery impacts. *Hydrobiologia, 471*(1–3), 1–12.

Foucault, M. (2005). *The order of things.* London, UK: Routledge.

Fredriksen, A. W., Løhre, M., Aarø, T., & Lorentzen, M. (2016, January 27). *Tillitsvalgte:—de ansatte er blitt overkjørt* [Union representative:—the employees are being disregarded]. E24. Retrieved October 15, 2021, from https://e24.no/energi/statoil/tillitsvalgte-de-ansatte-er-blitt-overkjoert/23313142

Friedman, A. L. (1977). *Industry and labour: Class struggle at work and monopoly capitalism.* London, UK: Macmillan.

Friedman, A. L., & Cornford, D. S. (1989). *Computer systems development: History organization and implementation.* Hoboken, NJ: John Wiley & Sons.

Frischmann, B. M. (2012). *Infrastructure: The social value of shared resources.* Oxford, UK: Oxford University Press.

Frodeman, R. (1995). Geological reasoning: Geology as an interpretive and historical science. *Geological Society of America Bulletin, 107*(8), 960–968.

Garfinkel, H. (1967). *What is ethnomethodology? Studies in ethnomethodology*. Upper Saddle River, NJ: Prentice Hall.

Garud, R., Jain, S., & Tuertscher, P. (2008). Incomplete by design and designing for incompleteness. *Organization Studies, 29*(3), 351–371.

Gawer, A. (Ed.). (2011). *Platforms, markets and innovation*. Cheltenham, UK: Edward Elgar.

Geels, F. W., & Schot, J. (2007). Typology of sociotechnical transition pathways. *Research Policy, 36*(3), 399–417.

Gerlitz, C., & Helmond, A. (2013). The like economy: Social buttons and the data-intensive web. *New Media & Society, 15*(8), 1348–1365.

Gillespie, T. (2010). The politics of "platforms." *New Media & Society, 12*(3), 347–364.

Gillespie, T. (2016). Algorithm. In B. Peters (Ed.), *Digital keywords: A vocabulary of information society and culture* (Vol. 2, pp. 18–30). Princeton, NJ: Princeton University Press.

Gitelman, L. (Ed.). (2013). *Raw data is an oxymoron*. Cambridge, MA: MIT Press.

Gjerde, K. L., & Fjæstad, K. (2013). Det meste er nord: Støres største satsing. *Internasjonal Politikk, 71*(3), 385–395.

Gjerstad, T. (2018, September 11). *Skjerper ressurskampen mot Russland* [Sharpens the conflict over resources with Russia]. Dagens Næringsliv. Retrieved October 15, 2021, from https://www.dn.no /politikk/olje/russland/barentshavet/skjerper-ressurskampen-mot-russland/2-1-417069

Glaser, V. L., Pollock, N., & D'Adderio, L. (2021). The biography of an algorithm: Performing algorithmic technologies in organizations. *Organization Theory*. Forthcoming.

Goodwin, C. (1994). Professional vision. *American Anthropologist, New Series, 96*(3), 606–633.

Graham, S., & Thrift, N. (2007). Out of order: Understanding repair and maintenance. *Theory, Culture & Society, 24*(3), 1–25.

Graham, T. (2018). Platforms and hyper-choice on the World Wide Web. *Big Data & Society, 5*(1). https://doi.org/10.1177/2053951718765878

Grisot, M., Kempton, A., Hagen, L., and Aanestad, M. (2019). Data-work for personalized care: Examining nurses' practices in remote monitoring of chronic patients. *Health Informatics Journal, 25*(3), 608–616.

Günther, W. A., Mehrizi, M. H. R., Huysman, M., & Feldberg, F. (2017). Debating big data: A literature review on realizing value from big data. *Journal of Strategic Information Systems, 26*, 191–209.

Hà, T. D., & Chow-White, P. A. (2021). The cancer multiple: Producing and translating genomic big data into oncology care. *Big Data & Society, 8*(1). https://doi.org/10.1177/2053951720978991

Haag, S., & Cummings, M. (2009). *Management information systems for the information age*. New York, NY: McGraw Hill.

Hacking, I. (1990). *The taming of chance*. Cambridge, UK: Cambridge University Press.

Handley, L. (2017, January 31). *Procter & Gamble chief marketer slams "crappy media supply chain,"* *urges marketers to act.* CNBC. Retrieved October 15, 2021, from https://www.cnbc.com/2017/01 /31/procter-gamble-chief-marketer-slams-crappy-media-supply-chain.html

Hanseth, O., & Bygstad, B. (2015). Flexible generification: ICT standardization strategies and service innovation in health care. *European Journal of Information Systems, 24*(6), 645–663.

Hanseth, O., Monteiro, E., & Hatling, M. (1996). Developing information infrastructure: The tension between standardization and flexibility. *Science, Technology, & Human Values, 21*(4), 407–426.

Harcourt, B. E. (2008). *Against prediction: Profiling, policing, and punishing in an actuarial age.* Chicago, IL: University of Chicago Press.

Harvey, F. (2021, May 18). No new oil, gas or coal development if world is to reach net zero by 2050, says world energy body. *The Guardian.* Retrieved October 15, 2021, from https://www .theguardian.com/environment/2021/may/18/no-new-investment-in-fossil-fuels-demands-top -energy-economist

Hauge, K. H., Blanchard, A., Andersen, G., Boland, R., Grøsvik, B. E., Howell, D., Meier, S., Olsen, E., & Vikebø, F. (2014). Inadequate risk assessments—a study on worst-case scenarios related to petroleum exploitation in the Lofoten area. *Marine Policy, 44*, 82–89.

Hecht, G. (2012). *Being nuclear: Africans and the global uranium trade.* Cambridge, MA: MIT Press.

Henfridsson, O., Nandhakumar, J., Scarbrough, H., & Panourgias, N. (2018). Recombination in the open-ended value landscape of digital innovation. *Information and Organization, 28*(2), 89–100.

Henke, C. R., & Sims, B. (2020). *Repairing infrastructures: The maintenance of materiality and power.* Cambridge, MA: MIT Press.

Hepsø, V., & Monteiro, E. (2021). From integrated operations to remote operations: Sociotechnical challenge for the oil and gas business. In N. L. Black (Ed.), *Proceedings of the 21st Congress of the International Ergonomics Association (IEA 2021): Systems and Macroergonomics* (Vol. 1, pp. 169–176). Cham, Switzerland: Springer Nature.

Hepsø, V., Monteiro, E., & Rolland, K. H. (2009). Ecologies of e-infrastructures. *Journal of the Association for Information Systems 10*(5): 2.

Hjelle, T. (2015). The role of strategies of practice-based learning for becoming a member of a community of practice. In B. X. Tung & R. H. Sprague Jr. (Eds.), *48th Hawaii International Conference on System Sciences* (pp. 3691–3700). Los Alamitos, CA: Institute of Electrical and Electronics Engineers Computer Society.

Hoeyer, K., Bauer, S., & Pickersgill, M. (2019). Datafication and accountability in public health: Introduction to a special issue. *Social Studies of Science, 49*(4), 459–475.

Holter, M., & Sleive, S. (2017, September 19). *The world's biggest wealth fund hits $1 trillion.* Bloomberg. Retrieved October 15, 2021, from https://www.bloomberg.com/news/articles/2017 -09-19/norway-wealth-fund-says-reached-1-trillion-in-value

Hoogendoorn, R., van Arerm, B., & Hoogendoom, S. (2014). Automated driving, traffic flow efficiency, and human factors: Literature review. *Transportation Research Record, 2422*(1), 113–120.

Hovland, M. (2018, January 17). *Installerer tusenvis av sensorer på havbunnen: Slik skal Statoil tømme Sverdrup* [Installing thousands of sensors on the seabed: This is how Statoil plans to empty Sverdrup]. E24. Retrieved October 15, 2021, from https://e24.no/olje-og-energi/i/ka9Pda/installerer-tusenvis-av-sensorer-paa-havbunnen-slik-skal-statoil-toemme-sverdrup

Hutchinson, B., Smart, A., Hanna, A., Denton, E., Greer, C., Kjartansson, O., Barnes, P., & Mitchell, M. (2021). Towards accountability for machine learning datasets: Practices from software engineering and infrastructure. In *Proceedings of the 2021 ACM Conference on Fairness, Accountability, and Transparency* (pp. 560–575). New York, NY: Association for Computing Machinery.

Igo, S. E. (2007). *The averaged American: Surveys, citizens, and the making of a mass public*. Cambridge, MA: Harvard University Press.

Ihde, D. (1995). *Postphenomenology: Essays in the postmodern context*. Evanston, IL: Northwestern University Press.

Ihde D. (1999). *Expanding hermeneutics: Visualism in science*. Evanston, IL: Northwestern University Press.

Inductive Automation. (2018, July 13). *What is IIoT? The industrial Internet of Things*. Retrieved October 15, 2021, from https://inductiveautomation.com/resources/article/what-is-iiot

Iyer, B., & Davenport, T. H. (2008). Reverse engineering Google's innovation machine. *Harvard Business Review, 86*(4), 58–68.

Jackson, S. (2014). Rethinking repair. In T. Gillespie, P. Boczkowski, & K. Foot (Eds.), *Media technologies: Essays on communication, materiality, and society* (pp. 221–239). Cambridge, MA: MIT Press.

Jarulaitis, G., & Monteiro, E. (2009). Cross-contextual use of integrated information systems. In *Proceedings of the European Conference on Information Systems 2009*. Association for Information Systems. http://aisel.aisnet.org/ecis2009/23

Jarulaitis, G., & Monteiro, E. (2010). Unity in multiplicity: Towards working enterprise systems. In *Proceedings of the European Conference on Information Systems 2010*. Association for Information Systems. http://aisel.aisnet.org/ecis2010/107

Jarvenpaa, S. L., Knoll, K., & Leidner, D. E. (1998). Is anybody out there? Antecedents of trust in global virtual teams. *Journal of Management Information Systems, 14*(4), 29–64.

Jensen, C. B., & Winthereik, B. R. (2013). *Monitoring movements in development aid: Recursive partnerships and infrastructures*. Cambridge, MA: MIT Press.

Johansen, E. N., & Kristensen, C. H. (2017, November 10). *Her blir det produsert gass for millionar utan folk om bord* [Natural gas is produced here for millions without people onboard]. NRK. Retrieved October 15, 2021, from https://www.nrk.no/vestland/no-blir-den-forste-plattforma-i-nordsjoen-ubemanna-1.13771938

Johnsen, W. (2020, January 24). *Spørsmål og svar om iskanten* [Questions and answers about the ice edge]. Norsk olje & gass. Retrieved March 20, 2020, from https://www.norskoljeoggass.no/om-oss/nyheter/2020/01/svar-om-iskanten/

Jones, M. (2019). What we talk about when we talk about (big) data. *Journal of Strategic Information Systems, 28*(1), 3–16.

Kallinikos, J. (2007). *The consequences of information: Institutional implications of technological change*. Cheltenham, UK: Edward Elgar.

Kallinikos, J., Aaltonen, A., & Marton, A. (2013). The ambivalent ontology of digital artifacts. *MIS Quarterly, 37*, 357–370.

Kallinikos, J., & Tempini, N. (2014). Patient data as medical facts: Social media practices as a foundation for medical knowledge creation. *Information Systems Research, 25*(4), 817–833.

Karasti, H., Baker, K., & Halkola, E. (2006). Enriching the notion of data curation in e-science: Data managing and information infrastructuring in the Long Term Ecological Research (LTER) network. *Computer Supported Cooperative Work, 15*(4), 321–358.

Kitchin, R. (2014). Big data, new epistemologies and paradigm shifts. *Big Data & Society, 1*(1), 1–12.

Kling, R. (Ed.). (1996). *Computerization and controversy: Value conflicts and social choices*. San Diego, CA: Academic Press.

Kling, R. (2000). Learning about information technologies and social change: The contribution of social informatics. *Information Society, 16*(3), 217–232.

Knol, M. (2013). Making ecosystem-based management operational: Integrated monitoring in Norway. *Maritime Studies, 12*(1), 5.

Knorr Cetina, K. (1999). *Epistemic cultures: How the sciences make knowledge*. Cambridge, MA: Harvard University Press.

Knorr Cetina, K. (2001). Objectual practice. In T. R. Schatzki, K. Knorr Cetina, & E. von Savigny (Eds.), *The practice turn in contemporary theory* (pp. 175–188). New York: Routledge.

Knorr Cetina, K. (2009). The synthetic situation: Interactionism for a global world. *Symbolic Interaction, 32*(1), 61–87.

Kongsvik, T., Johnsen, S., & Sklet, S. (2011). Safety climate and hydrocarbon leaks: An empirical contribution to the leading-lagging indicator discussion. *Journal of Loss Prevention in the Process Industries, 24*(4), 405–411.

Konkraft. (2018). *Konkurransekraft—norsk sokkel i endring. Utvalgets rapport februar 2018* [Competitiveness—Norwegian offshore is changing. The white paper report of February 2018] [White paper]. Retrieved October 15, 2021, from https://www.norskindustri.no/siteassets/dokumenter/rapporter-og-brosjyrer/2018-03-12-rapport-konkurransekraft_norsk-sokkel-i-endring.pdf

Kornberger, M., Pflueger, D., & Mouritsen, J. (2017). Evaluative infrastructures: Accounting for platform organization. *Accounting, Organizations and Society, 60*, 79–95.

Kotliar, D. M. (2020). Who gets to choose? On the socio-algorithmic construction of choice. *Science, Technology, & Human Values, 46*(2). https://doi.org/10.1177/0162243920925147

Krizhevsky, A., Sutskever, I., & Hinton, G. E. (2012). Imagenet classification with deep convolutional neural networks. In *Advances in neural information processing systems* (pp. 1097–1105). San Francisco, CA: Morgan Kaufmann.

Lahsen, M. (2005). Seductive simulations? Uncertainty distribution around climate models. *Social Studies of Science, 35*(6), 895–922.

Lamers, M., Pristupa, A., Amelung, B., & Knol, M. (2016). The changing role of environmental information in Arctic marine governance. *Current Opinion in Environmental Sustainability, 18*, 49–55.

LaPorte, T. R., & Consolini, P. M. (1991). Working in practice but not in theory: Theoretical challenges of high-reliability organizations. *Journal of Public Administration Research and Theory, 1*(1), 19–48.

Larkin, B. (2013). The politics and poetics of infrastructure. *Annual Review of Anthropology, 42*, 327–343.

Lasi, H., Fettke, P., Kemper, H. G., Feld, T., & Hoffmann, M. (2014). Industry 4.0. *Business & Information Systems Engineering, 6*(4), 239–242.

Latour, B. (1987). *Science in action: How to follow scientists and engineers through society*. Cambridge, MA: Harvard University Press.

Latour, B. (1992). Where are the missing masses? The sociology of a few mundane artifacts. In W. E. Bijker & J. Law (Eds.), *Shaping technology/building society: Studies in sociotechnical change* (pp. 225–258). Cambridge, MA: MIT Press.

Latour, B. (1993). *The pasteurization of France*. Cambridge, MA: Harvard University Press.

Latour, B. (1999). *Pandora's hope: Essays on the reality of science studies*. Cambridge, MA: Harvard University Press.

Lazer, D., Pentland, A., Adamic, L., Aral, S., Barabási, A. L., Brewer, D., Christakis, N., Contractor, N., Fowler, J., Gutmann, M., & Jebara, T. (2009). Computational social science: Obstacles and opportunities. *Science, 323*(5915), 721–723

LeCun, Y., Bengio, Y., & Hinton, G. (2015). Deep learning. *Nature, 521*(7553), 436–444.

Lehr, D., & Ohm, P. (2017). Playing with the data: What legal scholars should learn about machine learning. *University of California Davis Law Review, 51*, 653–717.

Le Masson, P., Hatchuel, A., Le Glatin, M., & Weil, B. (2019). Designing decisions in the unknown: A generative model. *European Management Review, 16*(2), 471–490.

Leonardi, P. M. (2010). From road to lab to math: The co-evolution of technological, regulatory, and organizational innovations for automotive crash testing. *Social Studies of Science, 40*(2), 43–274.

Leonardi, P. M. (2012). *Car crashes without cars: Lessons about simulation technology and organizational change from automotive design*. Cambridge, MA: MIT Press.

Leonardi, P. M. (2013). Theoretical foundations for the study of sociomateriality. *Information and Organization, 23*(2), 59–76.

Leonelli, S. (2014). What difference does quantity make? On the epistemology of big data in biology. *Big Data & Society, 1*(1). https://doi.org/10.1177/2053951714534395

Leonelli, S. (2019). Data—from objects to assets. *Nature, 574*(7778), 317–320.

Leonelli, S., Rappert, B., & Davis, G. (2017). Data shadows: Knowledge, openness, and absence. *Science, Technology & Human Values, 42*(2), 191–202.

Lie, E. (2017, May 6). *Ingen kan gjenta den norske oljesuksessen. Ikke engang vi selv* [Nobody can redo the Norwegian success with oil. Not even we]. Aftenposten. Retrieved October 15, 2021, from https://www.aftenposten.no/meninger/kommentar/i/zpQP1/ingen-kan-gjenta-den-norske-oljesuksessen-ikke-engang-vi-selv-eina

Lorentzen, M. (2015, May 11). *Høring om oljevirksomheten i Barentshavet: Oljebransjen advarer om at Russland kan komme oss i forkjøpet* [Hearing on oil business in the Barents Sea: The oil sector warns that Russia could get there before us]. E24. Retrieved October 15, 2021, from http://e24.no/energi/hoering-om-oljevirksomheten-i-barentshavet-oljebransjen-advarer-om-at-russland-kan-komme-oss-i-forkjoepet/23450809

Lusch, R. F., & Nambisan, S. (2015). Service innovation: A service-dominant logic perspective. *MIS Quarterly, 39*(1), 155–176.

Lycett, M. (2013). "Datafication": Making sense of (big) data in a complex world. *European Journal of Information Systems, 22*(4), 381–386.

Lyytinen, K., & Grover, V. (2017). Management misinformation systems: A time to revisit? *Journal of the Association for Information Systems, 18*(3), 2.

Mackenzie, A. (2017). *Machine learners: Archaeology of a data practice.* Cambridge, MA: MIT Press.

MacKenzie, D. (2006). Is economics performative? Option theory and the construction of derivatives markets. *Journal of the History of Economic Thought, 28*(1), 29–55.

MacKenzie, D., & Millo, Y. (2003). Constructing a market, performing theory: The historical sociology of a financial derivatives exchange. *American Journal of Sociology, 109*(1), 107–145.

March, J. G. (1991). Exploration and exploitation in organizational learning. *Organization Science, 2*(1), 71–87.

March, J. G. (1994). *Primer on decision making: How decisions happen.* New York, NY: Simon and Schuster.

Marcus, G. (2018). *Deep learning: A critical appraisal.* arXiv preprint arXiv:1801.00631

Markus, M. L. (2017). Datification, organizational strategy, and IS research: What's the score? *Journal of Strategic Information Systems, 26*(3), 233–241.

Marr, B. (2016). *Why everyone must get ready for the 4th industrial revolution.* Forbes Tech. Retrieved October 3, 2018, from https://www.forbes.com/sites/bernardmarr/2016/04/05/why-everyone-must-get-ready-for-4th-industrial-revolution/#74075eb83f90

Mateescu, A., & Eubanks, V. (2021, June 3). "Care bots" are on the rise and replacing human caregivers. *The Guardian.* Retrieved August 1, 2021, from https://www.theguardian.com/us-news/2021/jun/03/care-bots-on-the-rise-elder-care

McAfee, A., Brynjolfsson, E., Davenport, T. H., Patil, D. J., & Barton, D. (2012). Big data: The management revolution. *Harvard Business Review, 90*(10), 60–68.

Merry, S. E. (2009). Measuring the world: Indicators, human rights, and global governance. In D. S. Mathias, A. Roberts, & C. M. Vázquez (Eds.), *Proceedings of the ASIL Annual Meeting* (Vol. 103, pp. 239–243). Cambridge, UK: Cambridge University Press.

Meteorologisk Institutt [Norwegian Metrological Institute]. (2019, March 24). *100 måneder med temperatur over normalen på Svalbard* [100 months of temperatures above average at Svalbard]. Retrieved February 28, 2020, from https://www.met.no/nyhetsarkiv/100-maneder-med -temperatur-over-normalen-pa-svalbard

Mikalsen, M., & Monteiro, E. (2018). Data handling in knowledge infrastructures: A case study from oil exploration. In K. Karahalios, A. Monroy-Hernández, A. Lampinen, & G. Fitzpatrick (Eds.), *Proceedings of the ACM on Human-Computer Interaction* (Vol. 2, pp. 1–16). New York, NY: Association for Computing Machinery

Mikalsen, M., & Monteiro, E. (2021). Acting with inherently uncertain data: practices of data-centric knowing. *Journal of the AIS*. Forthcoming.

Miller, T. (2019). Explanation in artificial intelligence: Insights from the social sciences. *Artificial Intelligence, 267*, 1–38.

Millo, Y., & MacKenzie, D. (2009). The usefulness of inaccurate models: Towards an understanding of the emergence of financial risk management. *Accounting, Organizations and Society, 34*(5), 638–653.

Mindell, David A. (2015). *Our robots, ourselves: Robotics and the myths of autonomy.* New York, NY: Viking Adult.

Mitchell, T. (2011). *Carbon democracy: Political power in the age of oil.* London, UK: Verso Books.

Mnih, V., Kavukcuoglu, K., Silver, D., Rusu, A., Veness, J., Bellemare, M., & Graves, A. (2015). Human-level control through deep reinforcement learning. *Nature, 518*(7540), 529–533.

Mol, A. (2003). *The body multiple.* Durham, NC: Duke University Press.

Monteiro, E., Almklov, P., & Hepsø, V. (2012). Living in a sociomaterial world. In *Shaping the future of ICT research: Methods and approaches* (pp. 91–107). Heidelberg, Germany: Springer.

Monteiro, E., & Hanseth, O. (1996). Social shaping of information infrastructure: On being specific about the technology. In W. Orlikowski, G. Walsham, M. R. Jones, & J. deGross (Eds.), *Information technology and changes in organizational work* (pp. 325–343). Boston, MA: Springer.

Monteiro, E., Jarulaitis, G., & Hepsø, V. (2012b). The family resemblance of technologically mediated work practices. *Information and Organization, 22*(3), 169–187.

Monteiro, E., Østerlie, T., Parmiggiani, E., & Mikalsen, M. (2018). Quantifying quality: Towards a post-humanist perspective on sensemaking. In *Living with Monsters? Social Implications of Algorithmic Phenomena, Hybrid Agency, and the Performativity of Technology* (Vol. 543, pp. 48–63). IFIP Advances in Information and Communication Technology. Cham, Switzerland: Springer.

Monteiro, E., & Parmiggiani, E. (2019). Synthetic knowing: The politics of the Internet of Things. *MIS Quarterly, 43*(1), 167–184.

Monteiro, E., Pollock, N., Hanseth, O., & Williams, R. (2013). From artefacts to infrastructures. *Computer Supported Cooperative Work, 22*(4–6), 575–607.

Morgan, M. S. (2010). Travelling facts. In P. Howlett & M. S. Morgan (Eds.), *How well do facts travel? The dissemination of reliable knowledge* (pp. 3–39). Cambridge, UK: Cambridge University Press.

Muller, M. J., & Kuhn, S. (1993). Participatory design. *Communications of the ACM, 36*(6), 24–28.

Muniesa, F., Chabert, D., Ducrocq-Grondin, M., & Scott, S. V. (2011). Back-office intricacy: The description of financial objects in an investment bank. *Industrial and Corporate Change, 20*(4), 1189–1213.

National Commission on the BP *Deepwater Horizon* Oil Spill and Offshore Drilling. (2011). *Deep water: The Gulf oil disaster and the future of offshore drilling*. Nashville, TN: Perseus Distribution Digital.

Newell, A., & Simon, H. A. (1976). Computer science as empirical inquiry: Symbols and search. *Communications of the ACM, 19*(3), 113–126.

New York Times. (2018, October 4). Mark Zuckerberg testimony: Senators question Facebook's commitment to privacy. Retrieved October 15, 2021, from https://www.nytimes.com/2018/04/10/us/politics/mark-zuckerberg-testimony.html

Nilsen, J., & Stensvold, T. (2010, November 19). *Ptil med knusende Gullfaks-rapport* [The Petroleum Directorate with a devastating report about Gullfaks]. Teknisk Ukeblad. Retrieved October 15, 2021, from https://www.tu.no/artikler/ptil-med-knusende-gullfaks-rapport/252244

Nilsen, P. M. (2015, September 17). *Tar i bruk nytt anlegg til 19 milliarder* [Starting 19 billion new facility]. NRK. Retrieved October 15, 2021, from https://www.nrk.no/trondelag/tar-i-bruk-verdens-forste-undervannskompressor-1.12558093

Norsk olje & gass. (2017). *Fiskere i oljeberedskap* [Fishermen join in oil spill preparedness]. Retrieved November 1, 2017, from https://www.norskoljeoggass.no/en/News-archive/Miljo/Fishermen-to-join-oil-spill-preparedness-organization/.

Norsk Petroleum. (2021, October 12). *The income for the state*. Retrieved October 15, 2021, from https://www.norskpetroleum.no/okonomi/statens-inntekter/

Norwegian Climate and Pollution Agency. (2011). *Guidelines for offshore environmental monitoring—TA-2849/2011*. https://www.miljodirektoratet.no/globalassets/publikasjoner/klif2/publikasjoner/2849/ta2849.pdf

Norwegian Ministry of Climate and Environment. (2009). The Royal Norwegian Ministry of Climate and the Environment Act Relating to the Management of Biological, Geological and Landscape Diversity (Nature Diversity Act). No. LOV-2009-06-19-100. https://lovdata.no/dokument/NL/lov/2009-06-19-100

Norwegian Ministry of Climate and Environment. (2011). *First update to the integrated management of the marine environment of the Barents Sea and the sea areas off the Lofoten Islands* [White paper no. 10]. www.government.no

Norwegian Ministry of Oil and Energy. (2004). *Hjemfall* [White paper].

Norwegian Petroleum Directorate. (2019). Resource report 2019. Retrieved December 2021 from https://www.npd.no/en/facts/publications/reports2/resource-report/resource-report-2019/

Norwegian Petroleum Directorate. (2020, December 22). *Diskos*. Retrieved October 15, 2021, from https://www.npd.no/en/diskos/

Norwegian Petroleum Directorate. (2021, January 7). *About us*. Retrieved October 15, 2021, from https://www.npd.no/en/about-us/

NRK. (2021, April 21). *New oil discovery in the Norwegian Sea*. Retrieved October 15, 2021, from https://www.nrk.no/rogaland/nytt-oljefunn-i-norskehavet-1.15464577

Oborn, E., Barrett, M., & Davidson, E. (2011). Unity in diversity: Electronic patient record use in multidisciplinary practice. *Information Systems Research, 22*(3), 547–564.

O'Connell, J. (1993). Metrology: The creation of universality by the circulation of particulars. *Social Studies of Science, 23*(1), 129–173.

Oil & Gas Journal. (2008, June 23). Mud-pulse telemetry sees step-change improvement with oscillating shear valves. Retrieved October 15, 2021, from https://www.ogj.com/drilling-production/drilling-operations/article/17218455/mudpulse-telemetry-sees-stepchange-improvement-with-oscillating-shear-valves

Open Subsurface Data Universe. (2021). *The open group OSDU Forum enables the energy industry to develop transformational technology to support the world's changing energy needs*. Retrieved October 15, 2021, from https://osduforum.org/

Oreskes, N., Shrader-Frechette, K., & Belitz, K. (1994). Verification, validation, and confirmation of numerical models in the earth sciences. *Science, 263*(5147), 641–646.

Orlikowski, W. J. (2002). Knowing in practice: Enacting a collective capability in distributed organizing. *Organization Science, 13*(3), 249–273.

Orlikowski, W. J. (2006). Material knowing: The scaffolding of human knowledgeability. *European Journal of Information Systems, 15*(5), 460–466.

Orlikowski, W. J., & Scott, S. V. (2008). Sociomateriality: Challenging the separation of technology, work and organization. *Academy of Management Annals, 2*(1), 433–474.

Orlikowski, W. J., & Scott, S. V. (2016). Digital work: A research agenda. In B. Czarniawska (Ed.), *A research agenda for management and organization studies* (pp. 88–96). Cheltenham, UK: Edward Elgar.

Orr, J. E. (1996). *Talking about machines*. Ithaca, NJ: Cornell University Press.

OSPAR Commission. (2008). *Case reports for the OSPAR list of threatened and/or declining species and habitats*. Retrieved June 2015 from https://qsr2010.ospar.org/media/assessments/p00358_case_reports_species_and_habitats_2008.pdf

Østerlie, T., & Monteiro, E. (2020). Digital sand: The becoming of digital representations. *Information and Organization, 30*(1), 100275.

Ostrom, E. (1990). *Governing the commons: The evolution of institutions for collective action*. Cambridge, UK: Cambridge University Press.

Overby, E. (2008). Process virtualization theory and the impact of information technology. *Organization Science, 19*(2), 277–291.

Pacelli, V., & Azzollini, M. (2011). An artificial neural network approach for credit risk management. *Journal of Intelligent Learning Systems and Applications, 3*(2), 103.

Parker, G., & Van Alstyne, M. W. (2014). *Platform strategy*. London, UK: Palgrave Macmillan.

Parmiggiani, E., Monteiro, E., & Hepsø, V. (2015). The digital coral: Infrastructuring environmental monitoring. *Computer Supported Cooperative Work, 24*(5), 423–460.

Parmiggiani, E., Østerlie, T., & Almklov, P. G. (2021). In the backrooms of data science. *Journal of the Association for Information Systems*. Forthcoming.

Passi, S., & Jackson, S. J. (2018). Trust in data science: Collaboration, translation, and accountability in corporate data science projects. In K. Karahalios, K. Karahalios, A. Monroy-Hernández, A. Lampinen, & G. Fitzpatrick (Eds.), *Proceedings of the ACM on Human-Computer Interaction* (pp. 1–28). New York, NY: Association for Computing Machinery.

Pearl, J., & MacKenzie, D. (2018). *The book of why: The new science of cause and effect*. New York, NY: Basic Books.

Perrow, C. (2011). *Normal accidents*. Princeton, NJ: Princeton University Press.

Perry, J. S. (2017, May 22). *What is big data? More than volume, velocity and variety*. IBM Developer. Retrieved October 15, 2021, from https://developer.ibm.com/blogs/what-is-big-data-more-than -volume-velocity-and-variety/

Petroleumstilsynet [Petroleum Safety Authority Norway]. (2016). *Regulations relating to health, safety and the environment in the petroleum activities and at certain onshore facilities (the framework regulations)*. Retrieved October 15, 2021, from https://www.ptil.no/contentassets/bf47ce8fdec74 5f4a2c4ed073866f079/before-2013/rammeforskriften-2011_e.pdf

Petroleumstilsynet [Petroleum Safety Authority Norway]. (2018). *HSE effects of digitalization*. Retrieved October 15, 2021, from https://www.ptil.no/en/technical-competence/explore -technical-subjects/reports-from-projects/2018/hse-effects-of-digitalisation/

Pickering, A. (2010). *The mangle of practice: Time, agency, and science*. Chicago, IL: University of Chicago Press.

Plantin, J. C. (2019). Data cleaners for pristine datasets: Visibility and invisibility of data processors in social science. *Science, Technology, & Human Values, 44*(1), 52–73.

Plantin, J. C., Lagoze, C., Edwards, P. N., & Sandvig, C. (2018). Infrastructure studies meet platform studies in the age of Google and Facebook. *New Media & Society, 20*(1), 293–310.

Pollock, N., & Williams, R. (2010). The business of expectations: How promissor organizations shape technology and innovation. *Social Studies of Science, 40*(4), 525–548.

Pollock, N., & Williams, R. (2016). *How industry analysts shape the digital future*. Oxford, UK: Oxford University Press.

Pols, J., Willems, D., & Aanestad, M. (2019). Making sense with numbers. Unravelling ethico-psychological subjects in practices of self-quantification. *Sociology of Health & Illness, 41*, 98–115.

Poovey, M. (1998). *A history of the modern fact: Problems of knowledge in the sciences of wealth and society*. Chicago, IL: University of Chicago Press.

Porter, T. M. (1996). *Trust in numbers: The pursuit of objectivity in science and public life*. Princeton, NJ: Princeton University Press.

Power, M. (1997). *The audit society: Rituals of verification*. Oxford, UK: Oxford University Press.

Prentice, R. (2013). *Bodies in formation: An ethnography of anatomy and surgery education*. Durham, NC: Duke University Press.

Rahwan, I., Cebrian, M., Obradovich, N., Bongard, J., Bonnefon, J. F., Breazeal, C., Crandall, J. W., Christakis, N. A., Couzin, I. D., Jackson, M. O., & Jennings, N. R. (2019). Machine behaviour. *Nature, 568*(7753), 477–486.

Reinert, E. S. (2007). *How rich countries got rich and why poor countries stay poor*. London, UK: Constable & Robinson.

Reyna, S. P., & Behrends, A. (2011). The crazy curse and crude domination: Towards an anthropology of oil. In A. Behrends (Eds.), *Crude Domination* (pp. 3–29). New York, NY: Berghahn Books.

Rheinberger, H. J. (1997). *Toward a history of epistemic things: Synthesizing proteins in the test tube*. Stanford, CA: Stanford University Press.

Rheinberger, H. J. (2005). Gaston Bachelard and the notion of "phenomenotechnique." *Perspectives on Science, 13*(3), 313–328.

Ribes, D. (2019). STS, meet data science, once again. *Science, Technology, & Human Values, 44*(3), 514–539.

Ribes, D., & Finholt, T. A. (2009). The long now of technology infrastructure: Articulating tensions in development. *Journal of the Association for Information Systems, 10*(5), 375–398.

Ribes, D., & Polk, J. B. (2015). Organizing for ontological change: The kernel of an AIDS research infrastructure. *Social Studies of Science, 45*(2), 214–241.

Rogers, D. L. (2016). *The digital transformation playbook: Rethink your business for the digital age*. New York, NY: Columbia University Press.

Rommetveit, A., Duesund, J., Andersen, K. A. G., & Nøkling, A. (2017, July 1). *Iskantdramaet* [The drama on the ice edge]. Yr. Retrieved January 20, 2020, from https://www.yr.no/magasin /iskantdramaet-1.13560889#authors--expand

Rosendahl, T., & Hepsø, V. (2013). *Integrated operations in the oil and gas industry: Sustainability and capability development*. Hershey, PA: Business Science Reference.

Royal Society. (2012). *Science as an open enterprise*. http://royalsociety.org/policy/projects/science -public-enterprise/report/

Ryggvik, H. (2009). *Til siste olje*. Oslo, Norway: Aschehaug forlag.

Ryggvik, H. (2015). A short history of the Norwegian oil industry: From protected national champions to internationally competitive multinationals. *Business History Review, 89*(1), 3–41.

Ryggvik, H. (2018). Norwegian oil workers: From rebels to partners in the tripartite system. In T. Atabaki, E. Bini, & K. Ehsani (Eds.), *Working for oil: Comparative social histories of labor in the global oil industry* (pp. 99–130). London, UK: Palgrave Macmillan.

Salvesen, I. (2016, March 11). *Politikerne har bedt om det, og de har fått det* [The politicians asked for it, and they got it]. Dagens Næringsliv. Retrieved October 15, 2021, from https://www.dn.no/magasinet/dokumentar/klima/industri/jus/-politikerne-har-bedt-om-det-og-de-har-fatt-det/1-1-5598513

Sandbu, M. (2009, August 28). The Iraqi who saved Norway from oil. *Financial Times*. Retrieved October 15, 2021, from https://www.ft.com/content/99680a04-92a0-11de-b63b-00144feabdc0

Schlumberger. (1998). *Oilfield glossary*. Retrieved October 15, 2021, from https://glossary.oilfield.slb.com/

Schlumberger. (2021). *Petrel E & P software platform*. Retrieved October 15, 2021, from https://www.software.slb.com/products/petrel?entry=ad_google_sis_software_petrel&gclid=Cj0KCQjw5LbWB

Schüll, N. D. (2016). Data for life: Wearable technology and the design of self-care. *BioSocieties*, *11*(3), 317–333.

Seyfert, R., & Roberge, J. (Eds.). (2016). *Algorithmic cultures: Essays on meaning, performance and new technologies*. London, UK: Taylor & Francis.

Shafiee, K. (2018). *Machineries of oil: An infrastructural history of BP in Iran*. Cambridge, MA: MIT Press.

Shapin, S. (1995). Here and everywhere: Sociology of scientific knowledge. *Annual Review of Sociology*, *21*(1), 289–321.

Shapin, S. (2011). *A social history of truth: Civility and science in seventeenth-century England*. Chicago, IL: University of Chicago Press.

Shrestha, Y., Ben-Menahem, S., & Von Krogh, G. (2019). Organizational decision-making structures in the age of artificial intelligence. *California Management Review*, *61*(4), 66–83.

Silver, D., Huang, A., Maddison, C., Guez, A., Sifre, L., Van Den Driessche, G., & Schrittwieser, J. (2016). Mastering the game of Go with deep neural networks and tree search. *Nature*, *529*(7587), 484.

Silverstone, R., & Hirsch, E. (Eds.). (1994). *Consuming technologies*. London, UK: Taylor & Francis.

Simon, H. A. ([1969] 1996). *The sciences of the artificial* (3rd. ed.). Cambridge, MA: MIT Press.

Singh, D., Tripathi, G., & Jara, A. J. (2014). A survey of Internet-of-Things: Future vision, architecture, challenges and services. In *IEEE World Forum on Internet of Things* (pp. 287–292). Los Alamitos, CA: Institute of Electrical and Electronics Engineers Computer Society.

Sirius. (2021). *Sirius: Enabling digitalization in and beyond the oil and gas industry*. University of Oslo. Retrieved August 1, 2021, from https://sirius-labs.no/.

Sivarajah, U., Kamal, M. M., Irani, Z., & Weerakkody, V. (2017). Critical analysis of big data challenges and analytical methods. *Journal of Business Research*, *70*, 263–286.

Standards Norway. (2021). *NORSOK standards*. Retrieved October 15, 2021, from https://www.standard.no/en/sectors/energi-og-klima/petroleum/norsok-standards/#.YWlts3mxXrJ

Star, S. L. (1999). The ethnography of infrastructure. *American Behavioral Scientist*, *43*(3), 377–391.

Star, S. L. (2010). This is not a boundary object: Reflections on the origin of a concept. *Science, Technology, & Human Values*, *35*(5), 601–617.

Star, S. L., & Ruhleder, K. (1996). Steps toward an ecology of infrastructure: Design and access for large information spaces. *Information Systems Research*, *7*(1), 111–134.

Stark, D. (2011). *The sense of dissonance: Accounts of worth in economic life*. Princeton, NJ: Princeton University Press.

Stark, L. (2018). Algorithmic psychometrics and the scalable subject. *Social Studies of Science*, *48*(2), 204–231.

Statoil. (2017, November 9). *Norges første fjernstyrte plattform fra land* [Norway's first platform remotely operated from onshore]. Equinor. Retrieved October 15, 2021, from https://www .equinor.com/no/news/09nov2017-valemon-remote.html

Staudenmaier, J. (1997). Henry Ford's relationship to "Fordism": Ambiguity as a modality of technological resistance. In M. Bauer (Ed.), *Resistance to new technology: Nuclear power, informa-tion technology and biotechnology* (pp. 147–164). Cambridge, UK: Cambridge University Press.

Steinhardt, S. B., & Jackson, S. J. (2015). Anticipation work: Cultivating vision in collective prac-tice. In K. Karahalios, A. Monroy-Hernández, A. Lampinen, & G. Fitzpatrick (Eds.), *Proceedings of the 18th ACM Conference on Computer Supported Cooperative Work & Social Computing* (pp. 443–453). New York, NY: Association for Computing Machinery.

Støa, S. (2020, November 6). *Dette kan vi leve av etter oljen* [This we could live off after oil]. Sintef. Retrieved October 15, 2021, from https://www.sintef.no/siste-nytt/2020/dette-kan-vi-leve-av-etter -oljen/

Stortingsmelding. (1974). *Petroliumsvirksomhetes Plass i det Norske Samfunn* (The role of oil and gas activities in Norwegian society) [White paper]. Norwegian Ministry of Finance.

Suchman, L. A. (1987). *Plans and situated actions: The problem of human-machine communication*. Cambridge, UK: Cambridge University Press.

Suchman, L. A. (1993). Response to Vera and Simon's situated action: A symbolic interpretation. *Cognitive Science*, *17*(1), 71–75.

Suchman, L. A. (2006). *Human-machine reconfigurations: Plans and situated actions*. Cambridge, UK: Cambridge University Press.

Sugimoto, C. R., Ekbia, H. R., & Mattioli, M. (Eds.). (2016). *Big data is not a monolith*. Cam-bridge, MA: MIT Press.

Sundberg, M. (2010). Cultures of simulations vs. cultures of calculations? The development of simulation practices in meteorology and astrophysics. *Studies in History and Philosophy of Science Part B: Studies in History and Philosophy of Modern Physics*, *41*(3), 273–281.

Swanson, E. B., & Ramiller, N. C. (1997). The organizing vision in information systems innova-tion. *Organization Science*, *8*(5), 458–474.

Thune, T., Engen, O. A., & Wicken, O. (Eds.). (2018). *Petroleum industry transformations: Lessons from Norway and beyond.* London, UK: Routledge.

Timmermans, S., & Berg, M. (2010). *The gold standard: The challenge of evidence-based medicine and standardization in health care.* Philadelphia, PA: Temple University Press.

Tiwana, A. (2013). *Platform ecosystems: Aligning architecture, governance, and strategy.* Amsterdam, Netherlands: Elsevier.

Tuertscher, P., Garud, R., & Kumaraswamy, A. (2014). Justification and interlaced knowledge at ATLAS, CERN. *Organization Science, 25*(6), 1579–1608.

Turkle, S. (2009). *Simulation and its discontents.* Cambridge, MA: MIT Press.

US Department of the Interior. (2013). *Review of Shell's 2012 Alaska offshore oil and gas exploration program.* Retrieved March 1, 2015, from https://www.arctic-report.net/wp-content/uploads/2013/10/Department-of-the-Interior-Shell-Report.pdf

van den Broek, E., Sergeeva, A. V., & Huysman, M. (2021). When the machine meets the expert: An ethnography of developing AI for hiring. *MIS Quarterly.* Forthcoming.

van Heur, B., Leydesdorff, L., & Wyatt, S. (2013). Turning to ontology in STS? Turning to STS through "ontology." *Social Studies of Science, 43*(3), 341–362.

Vertesi, J. (2012). Seeing like a rover: Visualization, embodiment, and interaction on the Mars exploration rover mission. *Social Studies of Science, 42*(3), 393–414.

Vertesi, J. (2015). *Seeing like a rover: How robots, teams, and images craft knowledge of Mars.* Chicago, IL: University of Chicago Press.

Vial, G. (2019). Understanding digital transformation: A review and a research agenda. *Journal of Strategic Information Systems, 28*(2), 118–144.

Vieru, D., Rivard, S., & Dutot, V. (2014). Insights from a review of the literature on post-merger information systems/information technology integration. *International Business Research, 7*(7), 23–35.

von Krogh, G. (2018). Artificial intelligence in organizations: New opportunities for phenomenon-based theorizing. *Academy of Management Discoveries, 4*(4), 404–409.

Waardenburg, L., Sergeeva, A., & Huysman, M. (2018). Digitizing crime: How the use of predictive policing influences police work practices. In *34th European Group for Organizational Studies (EGOS) Colloquium: Surprise in and around Organizations: Journeys to the Unexpected.* Unpublished manuscript.

Watts, D. J. (2007). A twenty-first century science. *Nature, 445*(7127), 489–489.

Williams, R., & Pollock, N. (2012). Moving beyond the single site implementation study: How (and why) we should study the biography of packaged enterprise solutions. *Information Systems Research, 23*(1), 1–22.

Winner, L. (1978). *Autonomous technology: Technics-out-of-control as a theme in political thought.* Cambridge, MA: MIT Press.

Wong, B. K., & Monaco, J. A. (1995). Expert system applications in business: A review and analysis of the literature (1977–1993). *Information & Management, 29*(3), 141–152.

Wood, D., & Fels, J. (1992). *The power of maps*. New York, NY: Guilford Press.

Wylie, A. (2010). Archeological facts in transit: The "eminent mounds" of Central North America. In P. Howlett & M. S. Morgan (Eds.), *How well do facts travel? The dissemination of reliable knowledge* (pp. 301–322). Cambridge, UK: Cambridge University Press.

Wylie, A. (2017). How archaeological evidence bites back: Strategies for putting old data to work in new ways. *Science, Technology, & Human Values, 42*(2), 203–225.

Wylie, A., & Chapman, R. (2014). Learning from archaeological practice. In R. Chapman & A. Wylie (Eds.), *Material evidence: Learning from archaeological practice* (pp. 1–20). London, UK: Routledge.

Yoo, Y., Henfridsson, O., & Lyytinen, K. (2010). The new organizing logic of digital innovation: An agenda for information systems research. *Information Systems Research, 21*(4), 724–735.

Zittrain, J. L. (2006). The generative internet. *Harvard Law Review, 119*(7), 1974–2040.

Zuboff, S. (1988). *In the age of the smart machine*. New York, NY: Basic Books.

Zuboff, S. (2019). *The age of surveillance capitalism: The fight for a human future at the new frontier of power*. London, UK: Profile Books.

Index

Note: Illustrations and tables are indicated by page numbers in *italics*.

subsea technologies, *5, 5–6, 6,* 26, 38–40
Suchman, L. A., 74
swim bladder, 159–160
symbolic artificial intelligence, 12, 90–91
synthetic situation, 9, 171

Tempini, N., 165
3-D seismic cubes, 108
time-depth curve, 84
Timmermans, S., 173
trap, 26, *27*
Turkle, S., 119

uncertainty
    analogy and, 94
    in exploration, 95–98, *96*
    multiplicity and, 94
    prediction and, 93

Venus project, 150–154, *153, 155,* 158–161
Vertesi, J., 163
vitrinite reflectance, 100, 185n4
von Krogh, G., 13, 173

WDP. *See* wired drill pipe (WDP)
welfare technologies, 180, 187n1
well intervention, 54
well logging, 7, 19, 43, 51–52, *52,* 101, *101,*
    101–102
well maintenance, 54
well tie-ins, 102, *103,* 104, 108, 110
well-time, 84–85
Williams, R., 18, 93
wired drill pipe (WDP), 50
work, invisible, 20, 72–73, 76, 85, 92, 185n2
Wylie, A., 94, 174–175

Yoo, Y., 118

Zittrain, J. L., 118
Zuboff, S., 1, 8, 11, 68, 117, 176, 187n2
Zuckerberg, Mark, 68–69

Timothy Moss, *Conduits of Berlin: Remaking the City through Infrastructure, 1920–2020*

Claude Rosental, translated by Catherine Porter, *The Demonstration Society*

Blake Atwood, *Underground: The Secret Life of Videocassettes in Iran*

Huub Dijstelbloem, *Borders as Infrastructure: The Technopolitics of Border Control*

Dylan Mulvin, *Proxies: Standards and Their Media*

Eric Monteiro, *Digital Oil: Machineries of Knowing*